THE RISE
OF SELFISHNESS
IN AMERICA

THE RISE
OF SELFISHNESS
IN AMERICA

James Lincoln Collier

New York   Oxford
OXFORD UNIVERSITY PRESS
1991

Oxford University Press

Oxford New York Toronto
Delhi Bombay Calcutta Madras Karachi
Petaling Jaya Singapore Hong Kong Tokyo
Nairobi Dar es Salaam Cape Town
Melbourne Auckland

and associated companies in
Berlin Ibadan

Copyright © 1991 by James Lincoln Collier

Published by Oxford University Press, Inc.,
200 Madison Avenue, New York, New York 10016

Oxford is a registered trademark of Oxford University Press

Library of Congress Cataloging-in-Publication Data
Collier, James Lincoln, 1928–
The rise of selfishness in America /
James Lincoln Collier.
p. cm. Includes bibliographical references (p.   ) and index.
ISBN 0-19-505277-3
1. Self-interest—History—20th century. 2. United States—Moral
conditions—History—20th century. I. Title.
BJ1535.S4C65 1991
302.5'4—dc20 90-21127

2 4 6 8 9 7 5 3 1

Printed in the United States of America
on acid-free paper

For Ida, with love

Preface

This book grew out of my studies of early jazz, which led me to look into the American culture of the early twentieth century in which the new hot music developed. But, really, its genesis lies much further back, deep in my own boyhood. I was fortunate in experiencing at first hand the Victorian culture that our modern world sprang up in reaction to. This was possible because some of my forebears happened to live extremely long lives and carried into the second half of the twentieth century the ideas and attitudes they had been given in their youths. My paternal grandmother was born when Abraham Lincoln was President. She lived to be almost a hundred and three; I knew her well as an adult, and I understood her attitudes and beliefs.

More significantly, my mother's New England family was dominated by three maiden aunts who lived in a grand house on a hill, overlooking a row of what had once been the cottages of the workers their grandfather had employed. These three ladies clung to the Victorian lifestyle in which had been raised. True, by my time they took *The New Yorker* instead of *Punch;* cigarette smoking was permitted; and automobiles had replaced the carriages growing dusty in the carriage house.

But the essence did not change. Sex was not mentioned, liquor was forbidden (although one of the nephews kept a few bottles hidden in the basement; he once gave me a drink of pre-Prohibition whiskey—it was not very good.) Family was a primary concern—perhaps *the* primary concern. That meant that ancestors were spoken of a lot, but it also meant that the house was always there as a refuge for whoever of us needed it, however errant we had been. Church mattered; the aunts contributed regularly and went to service two or three times a week. Community was an obligation; they were involved with the library and youth organizations: my great-aunt Lydia was still in the 1950s meeting with the by-then graying members of the Girls' Friendly Society she had formed decades earlier. Culture, especially literature, was valued; various members of the household belonged to play-reading groups, book clubs, played the piano, went to the symphony. It was a disciplined—they would have said well-regulated—household, where order and decency prevailed.

I visited that house first when I was six months old and returned year after year to spend many hours and days there until I was well in my thirties, when the last of the aunts died. And I lived, so long as I was in that house, a Victorian life. As a boy I was read *The Peterkin Papers* and *The Children of the New Forest* before bed. I spent rainy afternoons curled up on the window seat in the second-floor hall listening to the grandfather clock tick and reading *A Child's History of England.* I went to Thanksgiving dinners with fifty relatives who ate three turkeys (one boiled and served with oyster sauce—it was very good), and after the meal we would all troop up into what had once been a billiard room and later a school room for the children of the house, to dig Victorian clothes out of horsehair trunks and play charades, three generations of family. I picked barley candies—yellow, translucent animals—from Christmas trees which sometimes were lit with real candles. I watched Fourth of July skyrockets explode over a lawn strung with Japanese lanterns, we children dressed in sailor suits in a row on the darkened lawn, eating lime sherbet and the deckle-edged sugar cookies which were a specialty of the house. I had tea with the minister—tea was served at four o'clock every afternoon, iced tea in the garden on nice days, hot tea in the sitting room otherwise. (There was of course a parlor filled with uncomfortable chairs, shelves of curios, and censorious ancestors staring down from the walls, but the aunts had enough sense not to spend much time there.)

I am by no means unaware that this comfortable life was purchased by the sweat of generations of mill hands working long hours for low pay. I know that these Victorian ladies lived far more constrained lives than most women would find acceptable today. And I eventually came to realize that behind some of those closed bedroom doors there was drug addiction, alcoholism, adultery.

But if these ladies regretted anything, they gave no clue. Once, in the course of a discussion of some point of history (these people loved history and read a great deal of it), my great-aunt Lydia said, "I'm so glad I was born a Victorian, because the ages before and after were so vulgar."

And she was right, at least partly. The Victorian Age has gotten an almost wholly bad press in the twentieth century. But as I hope this book will show, there was a lot of good along with the bad. That family feeling which I experienced, the concern for the people and the institutions of the town, was entirely genuine. It was simply taken for granted that people would give up things for each other, for ideals, for their community.

But if I experienced Victorianism, I also experienced the reaction to it. My mother was a feminist who started smoking at fourteen, bobbed her hair, rolled her stockings, went to college, and eventually worked on a newspaper. My father fled his Victorian home at sixteen to join the army, become a rover, itinerant worker, and eventually a writer. When my parents married they moved into Greenwich Village, where my mother's brother and his wife were at the heart of a circle of anti-Victorian bohemians which included Malcolm Cowley, Hart Crane, e.e. cummings, Eugene O'Neill, Edna St. Vincent Millay, Edmund Wilson, Allen and Caroline Tate, Matthew Josephson, Ford Maddox Ford, Kenneth Burke, and others. I grew up knowing some of these people slightly, some not at all. But some of them I knew very well. I knew what they had hoped for when they had charged out of their Victorian families; I knew what kind of world they were trying to make; I knew their victories and defeats, and what the battle cost them in terms of psychic disrepair.

Finally, in my own lifetime I have been able to see what happened next. As participant, reporter, and sometimes both, I have observed the major currents of the post-war world—the sexual revolution, the fight for black rights, the anti-war movement, the rise of the counterculture.

This book, then, is personal. I have, I hope, done the studying it calls for, but it is not simply researched—it is felt, for I have lived much of it.

Inevitably, I have had the help of many people in putting it together. I must first thank Christopher Collier and John L. Fell, who read the manuscript and offered many valuable criticisms and insights. H. Wiley Hitchcock was kind enough to give me his invaluable comments on American music. This might also be a good place to express my long-standing gratitude for the New York Public Library, one of the world's great treasure houses, where I have spent countless thousands of happy hours over many years.

Once again I have been lucky to have as an editor Sheldon Meyer, who, with his estimable staff have made my task as easy as it could be.

New York   J.L.C.
March 1990

Contents

THE RISE
OF SELFISHNESS
IN AMERICA

CHAPTER 1

The 18th-Century Debauch

We are in a crowded, smoky room. There is strongly rhythmic music playing, and a number of couples are doing dances with a good deal of twisting of the torso and swinging of the hips. Dress is casual and revealing, but meant to make a display. Nearly everybody is drinking—beer, wine, gin, whiskey are the most prominent, but some people are drinking less common spirits, like rum or brandy. There is a distinct smell of marijuana in the air; some people are using cocaine and perhaps other types of drugs. At one end of the room there is a sexual show in progress. From time to time couples slip out of the room, to reappear a half an hour or so later.

The foregoing scene could have taken place—and did innumerable times—in any of the so-called vice districts that existed in all of America's major cities and most of its smaller ones too in the years between, roughly, 1880 and 1910. They had become, by 1890 or so, a national institution of some importance. They consisted not merely of brothels, but of a number of adjunctive elements, such as saloons, gambling rooms, dance halls, theaters, and drug dens. They were inhabited by a floating population of prostitutes, pimps, gamblers, waiters, bartenders, and sports who did a little of everything; and they lived by catering to a vast crowd of males and at least a few females, drawn from all levels of society, who visited them to find forms of sensual pleasures denied them by the society in which they lived their diurnal lives. The vice districts were illicit and mostly illegal; but in the later days of the Victorian Age they existed everywhere in America, known to all, but spoken of in ordinary society rarely.

But the scene described, with its alcohol, drugs, music, and dance, could also take place today, not in an underworld vice district, but in the living rooms of millions of ordinary Americans in big cities and small towns, suburbs and farming villages all across the United States. What percentage of Americans has spent a Saturday night in such circumstances is hard to pin down; but, as we shall see later in this book, it is certainly a majority; and among the younger groups, a huge majority. There are of course differences of style between the vice district parlor house or saloon and the modern Saturday night party in Ashtabula or Larchmont. Then the music would have been syncopated pop tunes, ragtime, and in the latter days, the new jazz music; today it would be rock and its variants. Then the sexual show would have been live, performed on a small stage in a saloon or in the parlor of a brothel for a fee by entertainers who specialized in this kind of act; today it would be a film shown on a VCR or cable television. Then the dances would have been the schottische, the polka, and in the latter stages the new "trots," like the Turkey Trot and the Texas Tommy; today it would be any one of the dances that evolved out of the frug, the mashed potato, and the twist of the 1960s to accompany rock. Then there would have been more beer drunk and fewer mixed drinks than today; then there would have been more opium, along with cocaine and marijuana, and fewer of the prescription drugs used today. Then the couples slipping away would mainly have been prostitutes and their customers, rather than ordinary people involved with recreational sex.

But the essence has not changed. Then and now it was a gathering of people whose aim was a night of pleasure, of sensation seeking, a hedonistic fling when anything went and responsibility was kicked aside. What has changed is that what was once a hidden, forbidden activity engaged in furtively by American males from time to time as the occasion presented has become a common pattern which reaches through the entire society. Indeed, for many Americans it is almost a way of life, engaged in several times a week. What had once been unspeakable, at least among ordinary, decent people, has become commonplace—indeed accepted by many mainstream Americans.

The question this book asks is: How did we get from there to here? How in the course of about the sixty years from 1910 to 1970 did a morality that seemed fixed and permanent get stood on its head? Or to put it in broader terms, how did the United States turn from a social code in which self-restraint was a cardinal virtue to one in which self-gratification is a central idea, indeed ideal? How did we erect a moral code which has at its center the needs of the self—in which self-seeking is not merely condoned but actually urged upon us by philosophers, schools, television pundits, even recent governments? What happened?

In order to see how this change in America's social history came about, it will be necessary for us to back up a little and examine the society the change came out of. And I must immediately make the point that in generalizing about American culture I am leaving the special cases of blacks and American Indians out of the picture, at least for the most part. Both groups had, of course, profound effects on how the United States developed, and blacks especially have made a substantial contribution to American culture. But both groups were at least partially isolated from the mainstream of the culture—parallel subcultures which touched the main line at points, diverged at others.

My focus, then, will be on the broad, main line of development. In respect to that, there is a general feeling among Americans today that all ages before our own were "puritanical" and that the line of history has always been from greater repressiveness in personal behavior to greater freedom. This is not the case. Over history, at least in the so-called Western world, there has been a steady swing of the pendulum between times of greater and lesser restriction on sexuality, gambling, drinking, dancing, and other of the sensual pleasures. Or rather, to be more precise, there has been a succession of shifts in attitude toward self-gratification which were produced by different causes and took different shapes, but were generally of a piece in the sense that when one type of pleasure, such as sex, was anathematized, it was likely that others would be pressed down, too. For example, in the Italy of the late Renaissance, according to one authority, "debauchery and even crime had not the stigma which is attached to them today."[1] But slowly attitudes toward drunkenness, promiscuity, and vengeance murder hardened: the rakes of the 17th century would have found the Italy of the 19th century astonishing.

The relatively repressive Victorian Age of the 19th century had itself appeared in reaction to a quite different period that preceded it. That period, which is loosely known as the American Enlightenment, or simply the 18th century, was in some respects more like our own era than the intervening era of the Victorians.

For one thing, according to the noted historian Richard Hofstadter, in 1790 perhaps as many as 90 percent of Americans were "unchurched."[2] That is to say, they had no formal affiliation with any congregation, although of course the numbers actually attending church regularly were much larger. Americans of the 18th century, from the top of the social system to the bottom, were less religious than we are today.

For another, according to Norman H. Clark, who has studied the matter carefully, "In the early 1800s, the United States was an alcohol-soaked culture."[3] A careful study by Mark Lender and James Kirby Martin concluded, "While precise consumption figures are lacking, informed estimates suggest that by the 1790s an average American over fifteen years old drank just under six gallons of absolute alcohol a year . . . The comparable modern average is less than 2.9 gallons per capita."[4]

This may paint a picture that is somewhat misleading: actual drunkenness was generally frowned upon by most people, and certainly there

were many Americans who drank temperately. But according to Jack
Larkin, in an informative study of early American folkways, "Taverns
were surely the most widely accessible local institution of all. The great
majority of American men in every region were tavern-goers . . ."[5]
The drunken farmer staggering home from the tavern by moonlight
was a regular part of the nocturnal landscape.

The sexual appetites of these early Americans were similarly unfettered. The puritanism that obtained in much of 17th-century America
began to go by the boards early, and by the late 18th century existed
mainly as a memory. In this period, "Americans experienced a decline
in parental power over marriage choice, increased rates of premarital
pregnancy and illegitimacy, and the beginnings of conscious family lim-
itation," say John D'Emilio and Estelle B. Freedman in a recent study
of American sexuality.[6] According to another study, premarital pregnancy rates grew from under 10 percent in the 17th century to 30
percent in the second half of the 18th century.[7] Since a lot of women
who had sex before marriage did not become pregnant, and others
managed to conceal their condition, it is probable that at least 50 percent of American women of the time had sex before marriage, with
higher figures obtaining for men. Not until the middle of the 20th
century did premarital pregnancy rates approach those of that day.

And so it went, up and down the line. Gambling was a major leisure
occupation, the sport of roustabouts and gentlemen alike. "Most rural
taverns 'had their recess for gamblers,' " Larkin says.[8] Dancing was the
great social activity for people of all ages in a day before organized
sports. It was participated in by everybody, from the slave patting juba
in front of the cabin, to the elegant ladies and gentlemen at Mount
Vernon doing the rather vigorous reel.

The country was still, in many respects, a rude and uncivilized place.[9]
Few people had beds to themselves, and sometimes a bed was shared
with two or more, especially in the large families that were typical of
the day. They bathed infrequently and put on the same clothes morning after morning. They lived surrounded by manure: pig, sheep,
chicken, cow, horse dung was everywhere—in the city streets, casually
heaped up by farmhouses and barns, left to lie where it fell in fields
and on roadsides. Chamber pots were emptied into the streets, without
much regard for passersby. Outhouses were built conveniently close to
farmhouses to shorten treks through the cold in winter weather.

Broken windows, sagging doors, rotting clapboards went unrepaired
for months if not years, and houses were repainted infrequently. Junk—
the remains of broken tools, furniture, carts—was allowed to lie in
farmyards for years. Dogs, chickens, sheep, pigs roamed through
churches during services. Men and a great many women chewed tobacco, and gobbets of brown spit were everywhere, not merely on tav-

ern floors but on church floors as well. Many people ate their meals with knives only, and some depended primarily on their fingers.

Taken altogether, then, the reality of 18th-century America was at a considerable remove from the picture we have from our fourth grade textbooks of decorous ladies and gentlemen bowing and curtsying, and happy farmers at a barn-raising. A modern American transported back to a typical farm of the day would find himself assaulted by sights, smells, and rough behavior that would leave him dismayed. And as we shall shortly see, it was, by the end of the 18th century, beginning to dismay some of the people living amongst it at the moment.

CHAPTER 2

The Victorian Reaction

It is by no means easy to understand how and why a radical change in a culture occurs—and make no mistake about it, Victorianism was a radical movement, which not merely altered patterns of thought and habits of behavior of the 18th century, but in some cases turned them upside down. To simplify perhaps too drastically, the new Victorianism which began to flower in the first decades of the 19th century can be seen, at least in considerable part, as a reaction to the disorderly conduct of the 18th century. America was, after all, supposed to be a Christian nation, God-fearing and adhering to the ancient pieties.

Furthermore, their victory over the English in the great Revolution brought Americans to see themselves as a nation favored by God, for surely His hand was manifest in the triumph of their shabby, ill-clothed, and badly armed troops over the mighty British army. They had adopted for themselves the ideal of "plainness," in contrast to the corruption of the Old World, with its brutalized peasant masses, dissolute courts, incessant warfare, and grinding poverty.[1] This vision of the new nation did not square with the filth, licentiousness, and disorder Americans saw all around them.

In addition, the bloody excesses of the French Revolution had sickened people everywhere. Americans had had their own small version in Shays's Rebellion of 1786–87, when discontented mobs of farmers burnt courthouses and attacked the authorities. Shays's Rebellion made it even clearer that the society needed to be reordered.[2]

Furthermore, Victorianism cannot be seen as solely an American phenomenon, for elements of the new morality were appearing in Europe as well, as the 19th century came on. It was particularly evident in England, and historians today tend to see it as an "Anglo-American" movement. The historian Daniel Walker Howe says, "Victorianism was a transatlantic culture—though in the largest sense it was only an English-speaking subculture of Western civilization."[3]

Nor was it just "happening." It was a "concious ideal," as another historian, Stowe Persons, puts it.[4] By the early decades of the 19th century thinkers like Ralph Waldo Emerson in the United States and John Stuart Mill in England, were actively promoting the new morality, corresponding with each other, and reading each other's books. They came to constitute what David D. Hall calls "an avant garde isolated in the midst of Anglo-American culture," who shared "common goals as reformers."[5] Yet, while those and other factors were undoubtedly at work, the causes of Victorianism remain obscure.

The new movement had roots stretching back into the 18th century, and it was not finally buried, as we shall see, until after World War I. But its rule as the dominant mode in America ran from about 1830, when it began to be widely accepted, to the early years of the 1900s, when new ideas in art, politics, philosophy, and life itself were battering it into retreat.

It was built on the twin pillars of order and decency. I must immediately qualify this statement by saying that in recent years many historians, and indeed academics in general, have preferred to look at their subjects through the prism of class, race, and gender. In so doing they tend to see Victorianism as the morality of the new middle class, and they emphasize in their writings the patriarchal family they believe to be central to Victorianism, and the sharpening of the division between blue- and white-collar workers.

This tendency for academics to approach their work with issues of class, race, and gender foremost is immensely controversial in academic circles and elsewhere. Although I have my own views of the matter, for the purposes of this book it is not, I think, necessary for me to address this issue. There is no doubt that during the Victorian Age a prosperous middle class came into being at the expense of the largely immigrant working class.[6] All of this we shall see in due course.

But my interest here is in another aspect of the Victorian period, the broader morality of decency and order which came into being early in the century and obtained through most of it. This aspect of Victorianism was not confined to a new middle class that even by the end of the century constituted not more than a quarter of the population, but touched people from the top of the society to the bottom. The new gentility,[7] the uprush of religious fervor,[8] the anti-drink crusade, the control of sexual impulse[9]—these related movements were fermenting not merely in the new town houses of the rising middle class, but in the homes of the farmers and artisans who still, through most of the 19th century, made up the largest chunk of American society.

Victorianism was a revolution in thought, attitude, and manner which touched virtually every aspect of ordinary life. It reversed attitudes toward pleasure and self-gratification, especially as it concerned such matters of the senses as sex, drink, and dancing. It saw the rise of an aggressive woman's movement, the parent of today's feminism. It brought with it a rededication to Christian religion, to honor in human relations, to a general decency in manners and expression. Underlying everything was the idea that human beings could not live solely to gratify themselves, but must in every sphere of life—in politics, in philosophy, in social commerce—take into account the needs of the family, the community, the country, or even the whole of humankind. When choices were being made, not merely must others be thought of, they must be given first consideration.

Victorianism's central tenet was self-control or, as David D. Hall puts it, "discipline."[10] We can see this at work in a change in what it meant to be a gentleman. In the 18th century you were either born a gentleman or you were not. People might rise out of the common herd, as Benjamin Franklin did, but it was not often done.

The new idea of the gentleman was different. Through the early years of the Victorian Age it more and more came to be thought that a gentleman was what a gentleman did. He could be anybody, no matter how low-born, if he thought and acted in the right ways. Persons says, "It was no longer beneath the dignity of a gentleman to be skilled in the use of tools, or to sit down at the dinner table, as Emerson regularly did, with the field hands."[11] Robin Gilmour, in his study of the English gentleman of the Victorian period says, "The desire to become a gentleman was not just a snobbish aspiration out of one's class, but was also a desire to be a gentle man, to have a more civilized and decent life than violent society allowed for most of its members."[12] According to Emerson, "The gentleman is a man of truth, lord of his own actions, and expressing that lordship in his behavior, not in any manner dependent and servile either on persons, or opinions or possessions."[13] "Redemptive selflessness," was one goal of life for him.[14]

But it was not enough to purify oneself: the gentleman ought to encourage others to improve themselves, too. Leslie Stephen, a classic mid-Victorian English gentleman, and an important critic and thinker, said, "Even a novel should have a ruling thought . . . and the thought should be one which will help purify and sustain the mind by which it is assimilated."[15]

This new idea of the gentleman inevitably impinged upon the concept of the family. At an earlier time, Mintz and Kellogg say:

> it had been an almost unquestioned premise that the father, as head of the household, had a right to expect respect and obedience from his wife and children. A father's authority over his family, servants, and apprentices was simply one link in the great chain of being, the line of authority descending from God.[16]

But the new gentleman, believing as he did in placing the welfare of others before his own, could hardly act like this. According to Mintz and Kellogg, as the 19th century came on, "Within marriage the old ideals of patriarchal authority and strict wifely obedience were replaced by new ideals of mutual esteem, mutual friendship, mutual confidence."[17] The role of the wife changed. The growth of industry made the labor of wives and children on the family farms less necessary. Children were now an expense, not useful farmhands. Birth rates dropped, and women found themselves with more leisure time. People who supported the new morality saw quite clearly that if children could be raised to it, in time the whole of society would be transformed. But care had to be taken: running through Victorian literature, philosophy, social thought was the theme of the dangerous influences a child might fall under. Again and again "bad influences" were cited as the reason why people went wrong.

It followed that mothers must become the primary agents for raising good little boys and girls. They were in the home, while fathers were gone for long hours at work. More important, women were thought to be more refined, more kind and gentle, more "spiritual"; and as such they should take the lead in the raising of a new breed of people. "The romantic image of woman as more refined, and purer, higher and more delicate was potent and relevant . . . ," Stowe Persons says.[18] No longer would children be "broken" by authoritarian fathers, but maternal affection would become the "linchpin" of the socialization of the child, as Mary Ryan puts it.[19] The consequence was that a "new conception of motherhood as a special and uniquely important form of women's work emerged in the Revolutionary years and was finding substantial acceptance in the early nineteenth century America," Jack Larkin says. "Not only did Evangelical preachers charge mothers with preparing their children for salvation, but others maintained that the safety of the Republic itself, or, at the very least, success in the rapidly changing social and economic order were at stake."[20]

Indeed, the more refined woman was held responsible not only for raising the children but for uplifting her coarser husband as well. She was expected to set a "tone" for the home, one result of which was, as the century progressed, to force males to do their smoking and drinking elsewhere, in saloons and private clubs. The home must be pure and inviolate.

To Victorians, then, the nuclear family was of the utmost importance, an institution of vast significance in the social system. As work became less of a family enterprise, almost everything else became family-centered, resulting in a "cult of domesticity," as recent scholars term it. "In the evening, many families reassembled to read aloud from the Bible, novels, or family magazines . . . By the mid-19th century the family vacation had appeared, as did a series of new family-oriented celebrations, such as the birthday party, Christmas, and Thanksgiving."[21]

Not surprisingly, the new Victorianism brought a return to formal religious observance. According to Richard Hofstadter, the number of Americans who were on church membership rolls nearly doubled between 1800 and 1850.[22] Moreover, there was a tendency for the established, more decorous sects to take over from the rougher evangelical tradition which had dominated worship on the early frontier farmsteads. Hofstadter says that, for example, among the Methodists, "The passion for respectability was winning significant victories over the itinerating-evangelical, anti-intellectual heritage from the previous generation."[23]

The new ideas were, inevitably, felt in art. Just as the home and the church were required to improve people, so art was now not merely to provide aesthetic pleasure but to uplift people morally and spiritually. John R. Reed, in his study of Victorian literature, points out that a common theme in Victorian novels was the danger of pride and passion. Humility was a frequent ideal. "Early Victorian sentiments had assumed a human will driven by passions that must be restrained to harmonize with the divine plan . . ."[24] He adds, "Illness could function as an important moral agent for the Victorians because they believed in the value of human suffering . . . the world was viewed as a scene for trial and probation."[25] Or as Leslie Stephen had put it, the novel should "purify and sustain"—that is, help people resist temptation.

Painting, too, must be uplifting. The most celebrated, and most popular, of the Victorian painters were people like Thomas Couture and Alexandre Cabanel who painted vast scenes, frequently drawn from classical literature, which often told a story, and carried an explicit moral message.[26]

Even architecture was supposed to be uplifting. Henry Russell Hitchcock, writing about early Victorian architecture in England, points out that certain styles of church architecture became for some "an essentially religious crusade deeply imbued with values both ethical and sacramental."[27] Indeed, it is the opinion of the social historian Lawrence W. Levine that by the middle of the 19th century all of art had become holy—"sacralized."[28] By this he means that it could not be handled casually as it had been earlier in the century, but must be presented with dignity and respect—"museumized," if I may use that term—because it was now not meant merely to give pleasure but to teach morality and uplift the soul. Levine shows that, for example, in the earlier years of the 19th century, the plays of William Shakespeare were highly popular, and were often distorted to the point of burlesque to amuse audiences. By the latter part of the century, the famous plays were being presented with much more respect for the text and Shakespeare's intentions.[29]

Coupled with this interest in art was a more generalized concern for culture, which manifested itself in a vast industry of self-education. According to Ann Douglas in her study of Victorianism, the period saw a boom in "lending libraries, the lyceum lecture series, the periodicals, each relatively unknown in the eighteenth century but all-pervasive by the middle of the nineteenth . . ."[30]

Yet another thread in Victorianism was what is generally called gentility. As we have seen, manners among 18th-century Americans were as rough as the lives they lived. But now, as the 19th century came on, manners, along with morals, were being cleaned up. To quote Jack Larkin:

> . . . dogs, turkeys and geese were banished from worship, and more exacting standards both physical and social, of churchly decorum, were enforced. Congregations voted to repair broken windows, cushion pews and install stoves to make winter services more bearable. Chewing tobacco and noisy eating and spitting during services gradually came to a halt.[31]

People began tidying up their yards, digging refuse pits for garbage instead of tossing it casually into the weeds. They began to trim their lawns, keep their houses painted and in good repair, plant roses and lilacs. Table manners improved: even preachers began exhorting their flocks to "go to town and get you a set of chairs, knives, forks, cups and saucers."[32] The idea of the daily bath began to spread through a population which had previously viewed it as a hardship. Householders set aside separate rooms for sleeping, and increasingly it became the norm for people, except husbands and wives, to have their own beds. Functions which had once been public, or relatively so, became private.

This new gentility did not stem merely from an interest in cleanliness. Underneath was an understanding that good manners were a reflection of an attitude toward others: the person who ate nicely and was courteous was merely demonstrating his concern for the people around him. "Bad manners are merely selfishness expressed in tones and conduct. Good manners are charity in speech and action."[33]

By the late 1820s drunkenness also came under attack. Paul E. Johnson, in his study of Rochester, New York, a burgeoning new city at the moment, says: "The temperance question was non-existent in 1825. Three years later it was a middle-class obsession."[34] It has frequently been said that the fight against alcohol, a major battle in the 19th century, was led by the new industrialists who wanted sober workingmen turning up on time on Monday morning without hangovers, and this is undoubtedly to an extent true. But the initial attack on alcohol grew out of the religious revivals of the 1820s and 1830s. Drunkenness was not genteel: the gentleman advanced by Emerson and the others could hardly come rolling home at midnight singing coarse tavern songs; and the spiritually refined mother would not, certainly, spend her afternoons getting jingled on sherry with the ladies of the neighborhood. The new ideal for people simply could not include liquor. "It was not only the need for clear-headed calculation at work," Johnson says, "but the new ethos of bourgeois family life which drove businessmen away from the bottle."[35]

As a result, the consumption of liquor dropped through the Victorian Age: where 18th-century Americans were drinking that six gallons of absolute alcohol annually (of the fifteen-year-old and over population), the figure for the second half of the 19th century was less than half of that,[36] despite the influx of millions of Irish and Germans with fairly heavy drinking patterns. The 19th century, too, saw the birth of the prohibition movement, which resulted, in the 1920s, in the banning of alcohol altogether. As early as 1855 some thirty states had laws aimed at limiting drinking.[37] By the end of the century, "In the homes of the middle and upper lower classes, especially in the small towns, liquor of any description was seldom found, except a rise pint or so of whiskey or brandy, which was tucked away in the medicine chest and used only in emergencies. At their social functions, lemonade was the great standby."[38]

Smoking, too, was anathematized, especially for women. A man might enjoy his pipe or cigar, especially in the garden, but smoking was hardly appropriate for a delicate lady.

But if there is any aspect of human life that the word Victorianism brings to mind it is sex, and specifically the repression of the sexual impulse. The change in attitudes toward sex began quite early, and was related to a change in the way women were viewed. We can see it earliest in changes taking place in 18th-century English literature, which was of course widely read in the United States. These early novels were to a substantial extent built around illicit sex—basically about adultery, whoring, and the seduction of virgins. Among the most popular books of the era were Defoe's *The Fortunes and Misfortunes of Moll Flanders*, the story of a woman seeking to rise through her charms; Richardson's *Pamela*, which chronicled the efforts of a young woman to maintain her virginity until she could get the scoundrel to marry her; and Sterne's *Tristram Shandy*, a stew of oblique sexual jokes, allusions, and puns. These books reflected the 18th-century attitude that women were rather wanton, low-minded people. "Well into the eighteenth century, womanhood was associated with deviousness, sexual voraciousness, emotional inconstancy, and physical and intellectual inferiority," say Mintz and Kellogg.[39]

But even fairly early in the 18th century, some of the very authors who were writing these sexual tales were developing a new ideal of womanhood. Ian Watt, in his influential study of these early novelists, points out that as early as the beginning of the 1700s it was coming to be believed by a few people that men and women had different sexual constitutions. Men were robust, randy, and somewhat coarse; women were delicate, refined, and had only a moderate—and modest—interest in sex.[40] Why this view began to come forward at this particular moment is once again difficult to determine; it may have had more to do with wishful thinking on the part of males than anything else. In any case, by mid-century there existed a new stereotype of the heroine who "must be very young, very inexperienced, and so delicate in physical and mental constitution that she faints at any sexual advance," Watt says.[41] A consequence was that words and phrases which in any way suggested sex started to become taboo in the presence of women, at least in the literature of the period. Stockings, pregnancy, certainly undergarments would not be mentioned in front of ladies, especially young virgins. By the end of the century, according to Watt, the poet Samuel Taylor Coleridge was suggesting that such magazines as the very popular *Tatler* and *Spectator* sometimes contained terms "which might in our day, offend the delicacy of female ears and shock feminine susceptibility,"[42] the point being that an earlier day did not find such terms offensive.

But even those 18th-century novels depicting women as refined and delicate creatures were mainly about sex, however obliquely presented. In contrast, for the 19th-century novelist, sex simply did not exist. Writers like Twain, Dickens, and the rest of the classic novelists of the Victorian Age had to avoid the subject. People got married and had babies, but a reader could spend years wading through this vast heap of novels and never suspect that babies were not brought by the stork or found in the cabbage patch. Many writers complained, and Twain was noteworthy for writing sexual materials for private circulation among his friends, such as the celebrated "1601"; but nobody dared break the taboo publicly.

It is important to realize that for the Victorians the publishing of sexual materials was almost more significant than the act itself. Many Victorian males, even of the upper classes, routinely visited brothels; some kept mistresses; and even premarital intercourse was not wholly eliminated. But the rule against pornography could not be breached with impunity. The first serious obscenity laws in the United States came into being in 1842 when a ruling gave customs officials the power to stop the importation of obscene pictures.[43] Decade by decade the anti-pornography laws were strengthened, until by the 1890s almost anything suggestive could be prosecuted. The campaign against sexually explicit material was based on the Victorian notion that the minds of people were always open to dangerous "influences."[44] Rape, as well as murder, would be the result, an attitude that was quite explicit in the South, where whites assumed that a black male would rape any white woman he chanced upon if he were given the opportunity. As a consequence, it was important that the passions of working men be not influenced by the sight of too much decolletage, or an exposed ankle.[45]

The attack on sexual expression inevitably resulted in a reduction of premarital sexual intercourse. As we have seen, by the end of the 18th century, sex before marriage was a norm in many social groups, and premarital pregnancy rates had reached 30 percent. Through the first half of the 19th century these rates fell, until they reached 10 percent in 1860.[46] By that moment the idea that women were ethereal creatures above the demands of the flesh was a hundred years or more old and had flowed into every crevice and nook in the society. At the end of the century most young males were hag-ridden with guilt about their sexual impulses. By the late 19th and into the early 20th century, say D'Emilio and Freedman, "Most of the young men viewed their sexual behavior as a problem, a sign of moral weakness and a failure of manly self-control."[47] This really was it: a male was more of a man for the number of women he avoided having sex with rather than the number he bedded down. To a respectable Victorian, a good self-image was concomitant with self-denial, not self-expression.

This repression of sex and especially the public expression of sexuality touched everything. In dress, for example, Americans after the French Revolution briefly adopted the revealing Empire fashion. According to one authority, "The aim was to look as much as possible like the classic statues of Ancient Greece and Rome. To this end, clothing was reduced to a bare minimum, a single garment, usually linen, known as the shift, under the transparent gown."[48] As another authority says, "the eighteenth and nineteenth centuries met in the middle of a decade of undress."[49]

But by the early 1820s, a change was coming. "The skirt became fuller, longer, and heavier," says Valerie Steele, an authority on fashion history. "Despite all subsequent changes in the shape of the skirt and sleeves, the basic lines of Victorian fashion were set: The typical female silhouette was essentially formed by two cones—the long, full, structured skirt and the tailored, boned bodies—intersecting at a narrow and constricting waist."[50] Victorian dress was not without its sexual implications, Steele insists. Although the heavy dresses eliminated even the suggestion of female legs, the tightness of the skirt around the narrow waist tended to emphasize the shape of hip, thigh, and rump.[51] But the erotic display was covert, rather than overt, as was so much else about sex in the Victorian Age. Just as nudity must come in the guise of art, so the female body must be seen dimly, under heavy blanketing.

Divorces began to be granted to women whose husbands were excessively lusty. A demand for sex every night was considered too much, and divorce was almost automatic when a husband required sex from his wife when she was menstruating.[52]

Art, too, was affected. Early in the century the famous dictionary-maker Noah Webster expurgated the Bible to remove portions he thought were not fit to read in church, and in 1818, in England, Thomas Bowdler published expurgated versions of Shakespeare and incidentally gave his name to the language in the term "bowdlerize."[53] In dress, in print, in court, in church, and finally in the marriage bed, the Victorians did what they could to contain the sexual impulse, until by the end of the century the very idea of sex was no longer visible on the surface of the society.

But we must not let the Victorian obsession with the control of sex overshadow what Victorianism was really about—order and decency. Decency meant presenting yourself modestly, cleanly, and politely. It meant keeping a tidy home, however small and impoverished. It meant raising children to respect others, to obey the social strictures, to be polite and mannerly. And it meant, above all, restraining the passions, controlling the expression of the impulses: you did not fling yourself around on the dance floor, did not whoop it up in saloons or experiment with outré forms of sex.

Decency was private, personal; order was its public form. It meant turning up for work punctually and doing the job properly. It meant using accepted political forms, rather than rioting in the streets or stoning the police. It meant working through organizations—the Victorian period was the great age of the voluntary association—rather than running as a mob. And it meant the elimination of disorderly enterprises like saloons, the rough theater, the bawdy house. "An integrated concept of civic participation which tied religion, politics, work, and home into an organic ideological whole, is crucial to our understanding of Victorianism," says Elaine Tyler May.[54]

It is clear today that the Victorian ideal worked more to the benefit of the prosperous classes than it did to the laboring masses. The new industrial system, on which the prosperity of the white-collar class depended, required "repression of emotions and spontaneous impulses in favor of punctuality, order, cleanliness and devotion to duty . . . ," as Richard D. Brown puts it.[55] There is no doubt that many employers, especially, were aware of this and quite consciously promoted the Victorian ethic in order to keep the work force under control.

But to see Victorianism solely as a system for controlling the laboring masses, or women, or children, is to view it through the lens of contemporary ideologies. To the Victorians, looking back on the harsh and frequently primitive ways of the 18th century—a time when violence, drunkenness and dirt characterized the society from top to bottom—it seemed self-evident that the new ideal was better. Were not even work-

ing people happier when husbands brought home their wages, however small, rather than spend them in the saloons? Were they not more
content in homes, however modest, that were clean and tidy than aboil
with dirt and confusion? It seemed obvious to most Victorians that
everybody, even those at the bottom of the economic heap, benefited
from an orderly and decent society. It is simply true that many Victorians thoroughly believed in the ideal.

Few of them lived up to the ideal completely, and a great many of
them did not manage to live up to it at all. The Victorian Era was the
great age of the brothel; it was a time when the rapacious industrial
robber barons routinely used private armies to break strikes; it was a
time when corruption in government reached a peak; a time when possibilities for women were sharply constrained.

What is surprising, however, is not that so many Victorians failed to
meet the Victorian ideal but the extent to which they did meet it. Premarital pregnancy rates dropped sharply; alcoholic intake was down
two-thirds from the dizzying heights of the previous era; church attendance rose dramatically; homes, farms, and streets became cleaner; casual violence was curbed. My point—and this is crucial to my argument—is that, however large the gap between the Victorian ideal and
Victorian behavior, these people, as a society, set for themselves goals
of social concern, charity, self-control, a decent regard for the welfare
of others, a willingness to protect the weak. They may have failed, but
at least they were trying.

We have, in this century, looked back on Victorianism with scorn,
seeing it as a pious fraud. This is unfair, and inaccurate. The people
who made Victorianism were attempting—the word is critical—to impose on themselves the highest ideas of honor, self-sacrifice, human
dignity, decency toward others, fellowship, and concern for the welfare
of the community. As one of them, Edwin L. Godkin, a social theorist
of the day, put it, "The man of culture is the man who has formed his
ideals through labor and self-denial." He has learned "the art of doing
easily what you don't like to do."[56]

To modern ears the idea of teaching yourself to do what you do not
want to do will sound bizarre, even faintly psychotic. But in the imper-
fect world the Victorians found themselves in, it did seem to have its
uses.

CHAPTER 3

The Rising City and the Threat
to the Victorian Order

The world into which Victorianism was born was quite a different place from the one in which it died. The 19th century was a period of enormous upheaval in American society—and indeed elsewhere in what is now called "the industrial world." America began the Victorian Age as a rural nation, in which the majority of people lived and died on farms, most of them independent one-family farms; when the Victorian Age ended, the nation was dominated by the big industrial city—teeming with immigrants—whose tentacles, made of steel and copper, reached into nearly every corner of the country. The Victorians made the modern world in which we now live.

The America of the early 19th century was a place of isolated farms, tiny villages, small towns. Not more than 10 percent of the population lived in places that could even charitably be called towns.[1] Only 5 to 6 percent lived in centers of 2500 people or more. A small percentage lived in villages of a dozen houses clustered around a crossroads or at a ford in a river. Another small number lived in "towns" consisting of a main street of a dozen shops, a bank, a post office, the workshops of some artisans, and a few side streets of small houses.

The bulk of the American population lived on farms, and their situation was in some respects unique to human life. People for tens of thousands of years lived in small groups, either the nomadic hunting and gathering groups of at most forty individuals, which continued to exist down to this century; or the small peasant villages which began to develop about 10,000 years ago, and eventually became the standard way of life for most of humankind. These peasant villages almost always centered on a cluster of huts, lodges, communal dwellings, surrounded by gardens and groves of trees which might or might not be worked communally. The noted anthropologist Robert Redfield has suggested that as recently as the middle of the 20th century perhaps three-quarters of the human race still lived in such villages.[2]

The American situation was different. Because there was so much land available—in the 18th century it seemed almost infinite—almost anybody who was willing to work at it could have his own farm. These farms tended to be between 50 and 150 acres in size, with the house and barns located somewhere near the center of the area for the sake of convenience. The result was that a huge proportion of the American population did not live in villages, but was dotted thinly across the countryside. Redfield says that "while isolated homesteads appeared in early times, it was probably not until the settlement of the New World that they made their 'first appearance on a large scale.' "[3]

I must point out again that I am leaving both blacks and American Indians out of the picture in order to concentrate on the broad mainstream of the culture. As we go along, however, we should bear it in mind that there existed these other groups who were not exactly in the mainstream culture, and not exactly out of it, either.

The family units in which most whites lived at the beginning of the Victorian Era were extremely independent. Nearly everything needed was made at home. The women carded wool or cotton, spun it into yarn, wove it into cloth, and cut and sewed it into shirts, trousers, and dresses. One long, tiresome day each year they made enough candles to last for the next twelve months. These families built their own houses and barns, brewed their own beer, churned their own butter, pressed their own cheese, split their own shingles, hammered together their own carts. Frequently they did their own doctoring, using traditional home remedies of herbs and vegetables; sometimes they built their own coffins and conducted their own funeral services. They bought only a few things they could not make: plowshares, rifles, axe heads; and that was not often, as these things would last for years. Only when they were better off than most people did they own clocks, pewter dishes, drinking glasses. They told time by the sun, ate from wooden plates and bowls with fingers and knives, and drank from tin mugs.

What they needed from the outside world—the social system at large—was very little. This independence was made possible, in part, by the enormous abundance available to them outside their doorsteps. The woods were full of deer, turkey, pheasant; the lakes full of trout and bass; the oceans full of cod, clams, and crabs. Inevitably, it was a strong family-centered system, and it is not surprising that the Victorianism that developed on this base placed so much importance on the family. These people needed each other. A basic fact of life was that a single man or woman with children to care for could not survive: it took two adults to keep a home going, and it is therefore not surprising that at times widows and widowers remarried on the afternoon of the funeral.

In farm families everybody counted, down to the small child whose tiny fingers quickly became adept at picking berries and finding the eggs the hens laid in the straw.

The independence of the small farms should not be exaggerated. Many farmers were tied to the national economic system to the extent that they sold, or bartered, a certain amount of wheat, wool, corn, cotton, apples—or the cider, whiskey, or yarn made at home from these products—in order to buy the few manufactured goods they wanted. And of course the larger farms, like the great plantations of the antebellum South, were essentially what we would today call agri-business: growing cash crops which were shipped, in many cases, to distant parts of the world.

But the small farms on which the bulk of Americans lived were, by comparison with most other human enterprises elsewhere, astonishingly independent. They did need some things from the outside world, but they were few.

Families were bound tightly together by the economics of the farm in only partly tamed, open countryside. Mary P. Ryan, in her study of early 19th-century upstate New York, says: "The availability of relatively plentiful land at the turn of the century allowed fathers to provide nearby farms for their sons. At the same time, the primitive nature of the market discouraged conversion of family property into cash or fluid assets that would allow the family members to collect their separate inheritances and disperse. The same conditions made it practical to support aging parents in the households of one of their several sons . . . The corporate family economy wove together all ages and sexes in the incessant and inescapable interdependency of making a living on the new land."[4]

These farm families were bound in a kind of proximity that Americans today would find difficult to bear. Members of families were physically together almost continuously. Children usually slept two or three to a bed until they were adults, and perhaps beyond; people worked in cooperative teams at both ends of a long saw, tossing hay up into a loft, berrying, cleaning and salting pork and cod. They prayed together at night, walked to church together on Sundays, experienced birth and death together frequently. They lived this way in part because most households were large. Families often included a widowed sister, an orphaned niece, perhaps the hired man of country lore. Thirty-five percent of Americans lived in households of seven or more, two-thirds in households of five or more.[5]

It should not be thought that all these families were happy ones, living out lives taken from Currier and Ives prints. There were always brutal fathers, alcoholic mothers, wanton daughters, laggard sons. But even in troubled families people lived in a web of human relations that were tightly bonded and frequently lasted from birth to death. The terms mother, son, father, sister did not merely describe varieties of consanguinity: they were felt.

Families were embedded in small communities, the ties to which were almost as tight as they were inside the family. Ryan points out that "private life" in the modern sense did not really exist in the farm communities. In this case it was less a matter of economics than of the maintenance of good relations among neighbors. The home was not every man's castle: authorities, like ministers, were free to march in and out of houses as they saw fit. Churches, operating on the assumption that they had a duty to oversee the behavior of their members, regularly tried parishioners for sexual misbehavior, mistreatment of wives, and other offenses. Usually, if the guilty party confessed his error it was sufficient to put him back into good standing in the community. These farm communities, says Ryan, "tended to downplay a second conceptual building block of society, the category of the individual. The restriction of privacy within the corporate community was not conducive to the development of an independent, autonomous, egocentric self. Indeed, communal institutions tried systematically to suppress these personality traits."[6] And she adds, "The family economy was no less [sic] hospitable to individualism. Its organic interdependence did not permit of either self-reliance or solitary employment."[7]

It has been customary for historians to see 19th-century America as an age of rampant individualism, typified by the ambitious entrepreneur forging an industrial empire out of gall and cleverness; or the lone adventurer setting off for the gold fields of the West. But the relatively few strivers stand out in bold relief against a mass of farm people enmeshed in a web of family and community relationships in which subordination to the wishes of the group, rather than a spirited individualism, was the essential rule. Nineteenth-century America was individualistic compared with the peasant cultures of Asia or the villages of old Europe; but it was not individualistic as we understand the term today. Most people lived in communities of face-to-face relationships, where virtually everybody that touched your life was a known personality. People were attached to few strings pulled from long distances by faceless bureaucrats. For better or worse, you were stuck with your minister, your neighbor, your brothers and sisters. The cards dealt you would have to be played.

Furthermore, although some regional and ethnic differences existed, America of the early 19th century was far more of a piece than it would become by the end of the century. Once again leaving aside Indians and blacks, it was overwhelmingly Protestant, 85 percent English-speaking, 75 percent of British or Irish stock. As has been pointed out by the historian Maldwyn Allen Jones, it was this uniformity which enabled "a single set of social and political institutions to be established which later additions to the colonial population had been unable to alter basically."[8]

And it was this same uniformity which made it possible for the new Victorian idea to spread rapidly through the society in the first decades of the 19th century.

Yet even as it was becoming the dominant code of American society, there were at work in the nation movements that would eventually kill it. America, by 1850, was changing dramatically. The causes of that change are obvious and well known: the rise of industry; a wave of technological innovation, especially in the second half of the century; and an enormous influx of immigrants from Europe, Asia, and elsewhere. By the end of the century these three interwoven currents had turned the United States topsy-turvy, leaving it dominated no longer by the independent farmers in their island communities, but by gigantic industrial cities.

The first of these three forces which would so dramatically change American life was the rise of industry, built on technological innovation.

Even before the 18th century ended, a youth from England named Samuel Slater came to the United States with designs in his head for the cotton spinning machinery he had worked in the industrial Midlands.[9] He very quickly grew rich on a mass production system in which each worker—dozens of them lined up in front of identical machines—contributed one small step in the production of the final product—in Slater's case, cotton yarn. Very quickly other entrepreneurs applied the same principle to a host of other products—clocks, axes, carts, lanterns, cloth, and eventually whole suits of clothing. Industry after industry arose, driven first by the cheap and abundant water power that was available almost everywhere in the United States, and then by the steam engine that more and more took over as the century progressed.

The rise of industry changed the nature of farming. Very quickly farmers saw that instead of using the land to raise food for their families, they were better off devoting at least a portion of it to a cash crop: wheat, wool, cotton, grains. With cash they could buy much of what they formerly made at home and perhaps even put aside a small amount in the bargain.

This change in turn had the effect of throwing a lot of home labor out of work. Females, who spent an enormous amount of their time spinning and weaving, were especially becoming surplus labor in the face of factory-made yarn, cloth, and eventually clothing. The answer: send them into the spinning mills, use the wages they earned to buy the fabric they had formerly made. And once again there was likely to be money left over to purchase the niceties the Victorian ethic demanded: metal forks, machine-made chairs, painted plates.

Surprisingly, a lot of farm people—mostly young—did not object to going into the mills. The work lacked variety, and having to pace your-self to a clock rather than the sun and the stars was often annoying. But, after all, how much more monotonous was working a carding machine than haying under a blazing sun, or running a spinning wheel in an icy farmhouse?

The industrial system was, thus, self-feeding. The more people turned to manufactured goods, the more hands were needed in the mills. The rise of industry moved the country away from barter to a market economy based on cash. Many farmers, for a variety of reasons, could not cope with the new system, and went under. These displaced farmers had, perforce, to go into the mills. Between 1880 and 1890, for example, half the towns in Iowa and Illinois declined in population, even though the population of the nation as a whole was rising at an enormous rate.[10]

Nineteenth-century industrialism was fed by an enormous burst of technological invention, especially in the post-Civil War period. The application of the screw propeller to steam-driven vessels at mid-century very quickly led to steamships which could cross the Atlantic in two weeks, and then, by the 1890s, a week, with obvious effects on immigration and international trade. The Morse telegraph, which was demonstrated successfully in 1844, created a revolution in communication which had an enormous impact on journalism, and both national and international commerce. The decade of the 1870s alone saw the development or invention of the telephone, the incandescent lamp, and the phonograph. Each of these technological achievements echoed through the system to far-distant places. The incandescent bulb led to the wiring of whole cities, which in turn led to the application of the electric motor to dozens, then scores, and finally thousands of devices for home use. The creation of the telephone and the phonograph began the development of the sound reproduction on which the greatest portion of the huge American entertainment system continues to rest.

Invention bred invention, creating whole great industries based on the new technology: an automobile industry, an airplane industry, a phonograph industry, a farm machinery industry, a steel industry built on the new Bessemer process, and much more. It all happened in an enormous rush.

It could not have happened, at least not at the frantic pace at which it occurred, had it not been for another of those phenomena that were standing Victorianism on its head. That was immigration. The United States is, of course, a nation of immigrants; there has never been a year since the first Europeans settled in the wild new land that there has not been a significant group of new arrivals. But in the years immediately after the signing of the Constitution, the numbers of immigrants were relatively small.

The first big explosion of 19th-century immigration to America began in 1844 and lasted until the start of the Civil War in 1860. This wave, made up largely of Germans and Irish, was fueled in part by the 1848 political upheavals, and by the potato famine in Ireland, although the desire to improve one's lot was probably the basic cause of this population movement. Between 1845 and 1854 some three million Europeans emigrated to the United States, in that brief period adding 15 percent to the American population. The peak year was 1854, when 387,353 newcomers arrived.[11] From the high point, immigration rates dropped, to slow considerably during the 1860s, in part because of the unsettled conditions created by the American Civil War.

A second surge of immigration began about 1880, this time mainly from the east and south of Europe—Jews driven out of the Pale by Russian pogroms and oppressive legislation; Slavs, Italians, and others hoping to escape extremes of poverty and difficult political conditions. Although these "new" immigrants, as they came to be called, dominated the inpouring during the years from 1880 to 1920, the Irish and Germans continued to come, and even as late as World War I, "Germans still comprised the largest single nationality among the foreign born."[12]

This second wave of immigration peaked in 1907, when 1,285,349 newcomers arrived. Thereafter the flood gradually subsided, as a series of new laws put up ever higher walls against immigration. All told, 18.2 million immigrants entered the United States between 1890 and 1920, with some 9.4 million arriving in the years from 1906 to 1915 alone. Between 1899 and 1924, when the National Origins Act sealed off the flood, 3.8 million of the immigrants were Italians, 1.8 million were Jews, 1.4 million were Poles, and 1.3 million were Germans.[13] There were as well significant numbers of other nationalities, among them Czechs, Scandinavians, Turks, Armenians, and, of course, especially on the West Coast, large numbers of Chinese, Japanese, and other Asian groups.

This extraordinary influx was at first welcomed by Americans, by and large. There was in the country a "population hunger," a need to fill all that empty land with productive people. According to John Higham, a leading authority on immigration: "The general public shared the businessman's inclination to evaluate the newcomers in tangibly economic terms . . . The federal government smiled on the transatlantic influx and for a time toyed with schemes to assist it."[14] But by the 1880s numbers of Americans of the older stock were beginning to wonder if this flood of strangers was necessarily a good thing. Working people in particular saw that immigrants were being used to depress wages and were even shipped around the country in bulk to break strikes.[15]

The native-born middle class saw the foreign influx as bringing in alien manners and mores. In particular, many of the newcomers seemed to the old stock to be noisy, boisterous, over-emotional. Jack Larkin notes that New England farmers appeared stolid and chill, and he quotes a 19th-century writer who said that when two of them met, "their greeting might seem to a stranger gruff or surly, since the facial muscles were so inexpressive, while, in fact, they were on excellent terms." By contrast the Irish immigrants seemed "loud, boisterous and gesticulating. Their expressiveness made Anglo-Americans uncomfortable."[16] As early as 1857 the New York diarist George Templeton Strong, on seeing an Irish woman wailing over her dead husband, wrote in his diary, "Our Celtic fellow citizens are almost as remote from us in temperament and constitution as the Chinese."[17] And when Henry James returned to the United States after twenty-five years in England he discovered, in walking around his old haunts in Boston, that "the people before me were gross aliens to a man, and they were in serene and triumphant possession."[18]

The net result was to produce an increase of what historians call "nativism"—the idea of "America for Americans" which has had a long life. Bills to slow immigration were introduced into Congress regularly from 1882.[19] At first the business community, athirst for cheap labor, was able to beat back the attacks of the nativists; but gradually, decade by decade, the anti-immigration forces raised the walls, and by the early 1920s the great immigration was over. There was, however, more to it than that. Stowe Persons says: "Although legislation formally ended the era it is fairly clear now that the movement was terminating for natural reasons without the intervention of legislation. In America the frontier had disappeared by 1890; population was increasing faster than the demands made upon it by the industrial plant except in peak years; and the frequency and severity of depressions attendant upon the 'maturing' of the industrial economy all made the United States a much less attractive haven for the immigrant than formerly."[20]

But by the 1920s the immigrants had changed the American culture irrevocably. That much is clear. To say precisely *how* the culture was changed is not so easy. The superficialities are obvious: Americans began to eat in Chinese restaurants, adopt for general use Yiddish terms like "schlepp," and "maven," and dance to Polish mazurkas. But the immigrants brought with them more than just colorful costumes and novel cuisines. They had come out of cultures that had implanted deeply in them ways of thinking and feeling which were as much a part of them as their blood systems, and almost as resistant to change. This is not the place for a disquisition on how human beings acquire culture—a subject that is little understood in any case—but if there is anything we have learned from social science in this century it is how hard it is to change the view of life acquired at an early age. The immigrants were no different from anybody else in this, and they did not come to the United States with the intention of parting with something so basic as their belief systems. Indeed, it must be said that the bulk of them were badly educated—a substantial proportion were in fact illiterate[21]— and it is doubtful that they had any idea that many of their most cherished beliefs were considered abhorrent in the country they were coming to. Many, of course, were eager to make their ways in the new world they found themselves in; but few expected to have to do it at the cost of their own ways of thinking and feeling.

For one thing, a substantial proportion of the immigrants were "birds of passage," to use the term then current, who came with no intention of staying permanently. Immigrants from the Balkans were very likely to come for a brief period to save up money with which to start a business or buy a farm, and then return: some 66 percent of Romanians, for example, eventually went home.[22] Italians were also very transient: about 45 percent returned,[23] and a great many Italians crossed the Atlantic in the spring to work in the building trades made prosperous by the growth of the cities, and went home when winter came.[24] The Slavs, too, frequently came with the expectation of returning and were thus willing to take the tough mining jobs, where they could earn money faster, on the assumption that it was only for a few years.[25] Of course in the end, many of those who intended to go back stayed on, perhaps because they never quite amassed the little fortune they wanted, perhaps because they married and had children, perhaps out of simple lethargy. Yet, so long as any of these people expected to return, they saw little purpose in "Americanizing" themselves. If they could live embedded in a replica of the home culture, as most immigrants managed to do, why bother to learn English, to start businesses, form families, or in other ways melt into the American landscape?

But even those intending to stay were resistant to assimilation in many respects. The Jews, for example, who had run in fear of their lives, had no desire to return, and only 4.3 percent of them did so.[26] The Irish and Germans, who had left behind difficult political and economic conditions, tended to stay as well. And they wanted to keep their folkways. Alan Kraut insists that new "immigrants did not undergo their metamorphosis passively,"[27] and that most "had few contacts with members of other groups and even less with native-born Americans."[28] Says John Bodnar, author of a highly regarded recent study of immigrants, "They accepted change, but invariably sought to temper its thrust."[29]

We cannot, of course, see the immigrants as all of a piece. The cultural differences between Chinese and Irish, Jews and Italians, were almost unbridgeable, and even into recent times youthful descendants of these different groups risked a beating if they strayed onto the wrong turf. Categorizing any ethnic group is risky, but Alan Kraut has suggested that "Italian, Greek, Slavic and Oriental homes tended to be parent centered, where Jewish families were child centered; that Italians and Jews were more expressive, Oriental, Mexican and Slavic people tended to veil emotion."[30]

Yet if each group had its own beliefs and folkways they had come out of places in their cultures which were similar in some respects; and the pressures they felt on them in their adopted country were much the same for all. By and large the immigrants were religiously quite devout. The bulk of them were either Roman Catholics or Jews,[31] but most of the rest—the Asians, the German Protestants, the Greek Orthodox Catholics—had been raised in homes where religion was always present and taken seriously. For many their religions were more important to them than their nationalities. Jews of the Pale hardly knew what nationality they were, so often had the area been shifted between Poland, Russia, and for a time Lithuania; in practical fact, for most Jews their religion *was* their nationality. The same held true to an extent of Italians and Germans, who had come from nations only recently formed, and for whom Easter and their saints' days were more significant festivals than the national holidays.

The immigrants, however, came into a secular society in which religion had no legal standing and was a matter of personal choice. As we have seen, there was a turning to religion during the Victorian Age; but even then only 50 percent of Americans actually belonged to a church, although many more did attend church with some frequency. Old stock Americans did not always look to God or the priest for solutions to problems, but more often to the secular authorities or to their own individual efforts. Religion did not hold the place in American society that it did in the places where most of the immigrants came from. In the stetls of the Jewish Pale or the villages of Calabria it was hardly possible to be a member of society at all without accepting the authority of the synagogue or the church.

Furthermore, the majority of immigrants had been raised in peasant societies, tilling the soil and living in tightly enclosed groups. They had learned, the hard way, to be exceedingly suspicious of political authority. "Long and bitter experience with government officials and the large landholders they protected left the *contadini* with a cynical attitude toward all forms of authority other than the family," Alan Kraut says.[32] A contemporary report agreed that, "Government has meant little more [to the immigrant] than a means of oppression. He distrusts those in authority."[33] The legal powers back home seemed never to do anything for them but tax and hedge them in with rules and regulations. The consequence of all this was that in times of troubles they turned not to public officials, but to Father O'Brien and Rabbi Goldstein.

Old stock Americans, by contrast, were intensely political creatures, perhaps the most political the world had ever seen up to that time. The population was largely literate, and thousands of newspapers, representing every conceivable political position—and some that were not really even conceivable—circulated in the country. Americans debated issues at windy length in tavern harangues, and they could sit—or even stand—for two hours at a stretch listening to a spellbinding political orator. They supported their favorite candidates enthusiastically, if not always intelligently, and they turned out to vote. We can hardly pretend that the American electorate of the 19th century was always judicious in its choices; a lot of politicians got themselves elected by providing barrels of whiskey to their constituencies on polling day. But Americans looked to political solutions to problems, and they turned to lawyers more often than to religious leaders for help.

The immigrant cultures, with their religious bases, were inevitably conservative. Women and children were to obey their husbands and fathers, and everybody must respect Father O'Brien and Rabbi Goldstein. The immigrants did not, despite the sentimental stories written about them, see themselves essentially as potential Americans. They defined themselves, rather, in terms of their ethnic groups. They formed social clubs, religious organizations. Coming from closed social groups, they tended to look for collective solutions to problems, which served them well when they began to unionize the country. John Higham says: "In the cities immigrant politics was machine politics; a politics of loyalty, authority, reciprocal obligation and personal service . . . Its style of operations, therefore, was antithetical to American individualism."[34]

By contrast, people of the native stock were imbued with what has been called the "ideology of success." The idea, if not the actuality, was that anyone who worked hard enough could rise. "By the end of the nineteenth century," Alan Kraut says, "the ideology of success was well ensconced in American mythology. The novels of Horatio Alger became manuals for the young."[35] By contrast, the immigrants came, for the most part, from rigid societies in which birth ordained your place. In the United States it was possible for a boy to come off the farm and in time run for high public office, even to become President. The peasant tilling the landlord's farm in Ireland, Calabria, or Bavaria could hardly hope to rise to be bailiff, much less mayor of the local village. The immigrants did not think in terms of rising; their view of life was quite different. Kraut continues, "Most immigrants arriving after 1880 had never been exposed to the ideology of success.

"Even more important, new immigrants brought with them their own definitions of success, or the good life, quite different from those set forth in America's mobility ideology."[36] For many of them the family and the church were far more significant than becoming "successful." It was more important to them to see a good meal on the table and a decent contribution made on Sunday than it was to amass a little stock of capital with which to start a small business. Says Kraut, "Most new immigrants did not throw themselves wholeheartedly into the pursuit of wealth; they were equally as interested in structuring their lives to protect family and religious customs brought from their native countries." [37] Italians, for example, did not see education as a way upward; they were instead suspicious of the schools, which they feared would alienate the children from the family and break down parental authority. [38] For Italian fathers "work was important, but must not intrude on family life. There was little hunger for promotions or fear of loss of face if they became plasterers or plumbers rather than professionals or independent entrepreneurs." [39] Inevitably, Americans driven to move endlessly upwards were scornful of Italians who seemed lazy or without ambition. Once again, the immigrants were not all of a piece: Jews for the most part, for example, tended to be success-driven. But the majority of immigrants was less so.

These new immigrants were determined city-dwellers. Some had come from European cities; and many of the Jews had come either from ghettos of cities like Vilna or Warsaw, or the small-town stetls. But the majority of the immigrants had come off the land; and even many of those who came from cities had started on farms, and had made a move to a European city before migrating further to the United States. Nine out of ten Italians had been agricultural workers at home [40]— indeed nearly serfs tied to the land. [41] The Chinese who came to the United States were mostly "impoverished agricultural laborers from Toishan." [42] They had spent their days scratching poor livings from thin soil, and they saw no romance in the countryside, nor did they take pride in hacking their own farms out of the wilderness.

To be sure, especially in the earlier years when there was plenty of land to be had almost for nothing, a good many immigrants had been lured west or into places like Minnesota, to start farms. But the bulk of them saw the land only as a cruel master and wanted no part of it. [43] Far better to work in the mines and the sweatshops and take home your pay on Saturday, no matter how small, than to live menaced by drought, locusts, the landlord, and the tax collector.

Here again they differed from the old line Americans who, however quickly they were fleeing from the land, clung to a vision of life on the farm as ideal. Perhaps the most popular of all art works created in the United States during the second half of the 19th century were the famous Currier and Ives series of prints of American scenes. Some of them were of sporting events, and some were cityscapes; but the best-known were dozens depicting country scenes—the farmstead by the dirt road, the sleigh ride through the snowy fields, the boys and girls skating on the woodland pond, and the like. [44]

Despite the fact that by the turn of the century the city had come to dominate American life—or perhaps because of it—popular culture was chock-a-block with the romance of the country. The Irishman George M. Cohan might write "Give My Regards to Broadway," but a count of the ASCAP hit-song list for the first decade of this century reveals many more songs with rural than with city themes: "School Days," "In the Shade of the Old Apple Tree," "Shine on Harvest Moon," "From the Land of the Sky Blue Water," "Put on Your Old Grey Bonnet," "Down by the Old Mill Stream," and many more.[45] The prevailing opinion of the old stock was that the city was essentially bad—a noisy, dirty, and amoral place which contrasted poorly with the healthy outdoor life of the farm. "That the city was an illusive trap set carefully for country virgins was a notion strongly endorsed [in the pre-World War I period] by numerous respected educators, clerics and lawyers."[46]

Most important of all from the viewpoint of this book, few immigrants subscribed to the basic dicta of Victorianism. Richard D. Brown says: "toward time-thrift, education, and temperance values were at odds with most of the native population . . ."[47] Consider, for example, the today much-battered work ethic which was so important to the Victorian mind. The idea that work was a good thing had perfectly reasonable roots. In the early days of colonization, those people who worked almost incessantly survived and those who played did not. Later on, when life was more secure, it was possible for the family willing to do extra work to improve themselves—by acquiring and cultivating additional land, by making or growing marketable commodities like rum, timber, wheat. Hard work, in the America of the 18th and 19th centuries, often brought rewards, at times considerable ones.

Similarly, for Victorian Americans, "Repression of emotions and spontaneous impulses in favor of punctuality, order, cleanliness, and devotion to duty was a social necessity."[48] It was only through this kind of ordered behavior that people could improve their conditions, rise, establish themselves in the social system. Thrift, sobriety, and industry did actually pay off—if not always—in most cases to one degree or another.

But the immigrants saw life in another way. Indifferent to Victorian notions of success, which they believed to be unobtainable in any case, they wanted to enjoy their lives as much as possible through the warmth of associations with family and friends; by means of such public entertainment as they could afford; and through drinking, and dancing in the saloons, concert gardens, and taverns they created for themselves in their own neighborhoods. The work ethic meant little to them: they had come out of cultures where work got you nothing but calluses and a sore back.

The Germans in particular "insisted belligerently on their right to amusements that shocked the censorious—to card playing, to beer gardens, to Sunday frolics, and when the temperance issue revived in the seventies, the *Chicago Tribune* thought enforcement of the Sunday closing law necessary to prevent 'the German Conquest' of the city."[49] Another observer says, "The Sunday parades [of the Germans], the bands, the beer drinking—all seemed to Europeanize and destroy the sanctity of that special day."[50]

It was much the same for other groups. Irish males came out of a culture in which whiskey played an important role. In Slavic neighborhoods "tavern-keepers were usually the most prosperous."[51] One contemporary study reported that 669 of 865 liquor license applications processed during a certain period in Philadelphia were submitted by naturalized citizens. Germans, the same report said, "largely controlled the brewing interests."[52] Not all immigrant groups were afloat in alcohol: drinking rates among the Italians and the Jews were lower than they were for the old stock.[53] But even in these groups there was none of the anti-liquor fever which was rising in the Victorian Age.

The immigrants also brought with them a tradition of public dancing. As we have seen, among the Victorian Americans there was considerable ambivalence about the moral worth of dancing. Some people gave dances in their homes, but the more religious did not. In any case, nobody but low types danced in public; for a woman to do so was almost, in the eyes of the middle class at least, to brand herself as promiscuous. But many of the immigrant groups had long made dancing a regular part of holidays and celebrations. Dancing was almost a ritual part of weddings among the Jews, Italians, Poles, and other groups, and even today the words "a Polish wedding" or "an Italian feast" conjure up images of music and very energetic dancing. It was the immigrants, more than anybody, who created in the late 19th and the early 20th century the public dance hall and the concert garden, where dancing along with music, food, and drink were staple pleasures for Saturday night and Sunday afternoon.

Nor did the immigrants anathematize sex as the Victorians had come to do by the latter decades of the 19th century. Coming from conservative cultures dominated by religion, the immigrants tended to demand virginity for women before marriage, and a wife who committed adultery, or even permitted the suspicion of it to arise, could expect a severe beating. But broadly speaking, attitudes toward sexuality in immigrant groups were generally more liberal than they were among the old stock. It is difficult to put figures to this; but a 1970 study by the Kinsey Institute concluded that Jews were among the most liberal of religious groups in their view of sex; and that surprisingly, the Catholics tended to be about as liberal as the more liberal Protestant groups.[54] Jews and Catholics constituted the bulk of the immigrants; and as they could not have acquired liberal sexual views from the Victorian society around them, it is fair to assume that they brought this attitude with them.

The immigrants, then, were bringing to the United States an array of habits, attitudes, and folkways that conflicted, at times dramatically, with the prevailing American patterns of thought and behavior. They were, in sum, resolutely anti-Victorian in almost every respect. They did not believe in discipline, punctuality, sobriety—the order and decency of the Victorian ethic. They wanted instead to live as expressively as they could. In what spare time they could snatch from their jobs and family obligations they wanted to drink, to dance, to gamble, to have fun. It is hardly surprising, therefore, that the people of the old stock were appalled by their behavior. It seemed to them that the newcomers were intent upon destroying the decent and orderly society that they of the old stock were trying so hard to build and maintain.

The answer they proposed was to "Americanize" the newcomers, through the settlement houses that appeared in the 1890s in the cities of the Northeast, through the school, through night classes in the English language and American history. "As both settlement and charity workers understood it, 'victory' in this cultural confrontation was the achievement of a cultural monism in which the immigrant hordes surrendered their distinctive traits, accepted middle class values, and patterned their institutions on the American example," writes Paul W. McBridge.[55] As the American middle class saw it, everything would be all right if only the immigrants would become "more like us."

It is easy to sneer at this position as elitist and condescending, which it no doubt was. But it is also easy to understand the resentment of people who suddenly found the town, the city, the neighborhood they had grown up in, and which they thought of as theirs, chockablock with strange and noisy people who refused to accept the established values of the original residents, some of whose families had been there for generations. Many of the people of the old stock felt dispossessed, and it is hardly surprising that they would want to reimpose the values they had grown up with: after all, their values meant as much to them as other values did to the immigrants.

But it was naïve for the old stock to think that the newcomers could abandon the ways of their cultures as easily as they could adopt new styles of dress. "The majority of new immigrants found themselves being pulled in both directions while they tried to design a compromise between total absorption and total alienation," says Alan Kraut.[56]

In sum, although it is certainly true that some immigrants did go to night school to learn English, did consciously study American folkways, the bulk of them to one degree or another resisted change. And they were bringing to the United States an approach to life that was, paradoxically, less optimistic and more hedonistic than the Victorianism of the old stock. They did not believe that the point of life was to rise through hard work. They accepted a measure of hard times as inevitable; they wanted only to temper their difficulties with what pleasure they could get. They were, inevitably, living by a different set of rules from the people who had come before them.

CHAPTER 4

The New Class System

In the last decades of the 19th century, the process which was bringing into being the industrial city sped up. According to Ray Ginger, one student of the period, "Manufacturing production as a whole in 1892 was 2¼ times the level of 1877." In that time the number of industrial workers in the country more than doubled, while the agricultural work force increased by less than 50 percent.[1] Technologies gave us the automobile, the electric motor, the airplane. The immigrants continued to pour in. "By the turn of the century," Alan Kraut says, "the new immigrants were becoming the chief source of labor in almost every area of industrial production."[2] They "furnished the bulk of the manpower" for the new industrial machine, John Higham agrees.[3] They were essential, for it is unlikely that the old stock Americans coming off the farms could have or would have supplied enough labor. Kraut says, "Most immigrants . . . did not perceive themselves to be victimized by the new industrial order as many American born workers did." Conditions for them back home had been even worse.

No society could undergo such dramatic changes in so short a time without experiencing grave stresses. Particularly during the economic crisis of the 1890s, but at other times as well, the country saw epidemics of industrial strife, in which troops were called out, workers shot dead, factories bombed. Corruption in government was rife, and big business demonstrated an indifference to the common good rarely matched. Richard Hofstadter said: "In business and politics the captains of industry did their work boldly, blandly, and cynically. Exploiting workers and milking farmers, bribing congressmen, buying legislatures, spying upon competitors, hiring armed guards, dynamiting property, using threats and intrigue and force, they made a mockery of the ideals of the simple gentry who imagined that the nation's development could take place with dignity and restraint under the regime of laissez-faire."[4]

But despite everything—industrial strife, corruption, economic debacle—the United States not merely survived, but grew fat. And by the turn of the century, or thereabouts, the combined forces of technological innovation, immigration, and urbanization had created something new to the United States—the industrial city. America was not the first nation to bring forth this institution, which has come to dominate the modern world: London and Paris were both modern cities by the mid-Victorian Age.

But the United States, with its exploding population, its vast natural resources, and its tradition of entrepreneurial self-enterprise, was by the last decades of the 19th century catching up.

In 1860, some 6.2 million Americans lived in areas defined as "urban," while 25.2 million lived in rural areas. But this is misleading, for any place with over 2500 people was termed urban, which put a substantial number of farming centers in that category. Under this definition, in any case, 20 percent were urban. By 1870 the population of the country had increased by about 21 percent, but the urban population had grown by 50 percent. In the next ten years the population grew by about 30 percent, while the urban population increased by 40 percent. And in the decade of the 1880s, while the national population went up about 22 percent, the urban population increased by over a third: now, in 1890 over a third of Americans were city-dwellers. By 1910 it was 40 percent, and somewhere around the time of the start of World War I the rural-urban balance tipped, with more than half of the population living in cities.[5]

Overall, between the end of the Civil War, and the beginning of World War I, a period of about two generations, the urban population of the United States increased seven-fold, while the rural population just doubled.

This huge and incredibly rapid growth of the cities was by no means built solely on immigration. For example, in the first decade of this century, some 8.7 million immigrants arrived.[6] The population of the country as a whole, however, grew by about 15 million, and the urban population increased by 12 million. If we assume that while the newcomers were flooding in, a large group was also returning home, it becomes clear that even in this decade in which immigration peaked, half of the new urban population (leaving aside some natural increase) was due to the flooding in of Americans from farms and villages. This is borne out by figures produced by Robert S. Lynd and Helen Merrell Lynd in their famous study of Muncie, Indiana, which they termed "Middletown," in the 1920s. According to the Lynds' survey, half of the working-class wives in what was by then a highly industrialized city had been born on farms, most of them presumably between about 1870 and 1900.[7] The Lynds wrote, "The Middletown working man is American born of American parents. He lives on a middle Western farm, has moved in from the farm, or his father's family moved to town from a farm."[8] It was not just in Muncie, however: in New York City, between 1870 and 1890 the percentage of foreign-born actually dropped slightly, from 38 to 36 percent,[9] as more people came to the city from the farm than from abroad.

This fact, that the new urban population was divided roughly equally between foreigners and old stock Americans from off the farms, had tremendous consequences for the developing urban society in which most of us live today, for it provided the base for the creation, beginning in about 1830, of a new middle class whose ideals, aspirations, and moral system would, for better or worse, come to dominate American society. It is astonishing how little this new class has been studied: the literature on the immigrants has been voluminous, with dozens of books devoted to the experiences of virtually every ethnic group to come to the United States, no matter how small; but only relatively recently have historians begun to produce major studies of the middle class that developed in the nineteenth century.

Through most of the earlier centuries of the country's existence, America was an amazingly egalitarian place, once again leaving aside blacks and Indians. There was on the top a tiny handful of wealthy and educated people who by birthright played most of the important political and social roles. These were the Madisons, Jeffersons, Quincys, Langdons, and the rest—many of them trained in the law—who constituted an elite whose right to lead was accepted not only by themselves but by the mass of people everywhere.[10] At the bottom of society was another small handful of the dispossessed—day laborers, prostitutes, criminals,—many of them residing in cities. Taken together, these two small minorities at the top and the bottom hardly amounted to 10 percent of the population. The remaining 90 percent of Americans were farmers and artisans. They were amazingly alike in the value of the property they owned, their incomes, the kind of work they did, and their social status. To be sure, there were degrees: some people prospered more than others, some ate from pewter and drank from glass while others used wooden trenchers and wooden mugs; some had clocks while others told time from the sun; some owned saw and flour mills, while others scratched a bare living from rocky soil on mountainsides.

In particular, historians have pointed to the existence of a group of artisans and small shopkeepers who were sometimes termed "the middling sort." These people, who ran tailor shops, made and sold hats, ran cooperages and smithies, according to some historians, constituted an incipient middle class. But in fact, neither in wealth nor in lifestyle did they differ essentially from most farm families. Susan E. Hirsch has shown that in Newark artisans and farmers were linked in people's minds

as coming from comparable circumstances and having related interests.[11] Blumin has shown that many, if not most, people "of the middling sort" in fact had marginal incomes, and Hirsch concludes: "The
artisan class was part of a preindustrial American class structure and
was not comparable to any class in our later industrial society."[12] This
class of artisans and small shopkeepers did not flow into a new middle
class; instead, it simply disappeared as the nation industrialized.

Thus, while inequalities certainly did exist in the earlier time, they
were not significant enough among the bulk of Americans to divide the
nation's central population into easily described classes. Hofstadter says,
"Up to about 1870, the United States was a nation with a rather broad
diffusion of wealth, status, and power."[13]

A group of researchers who returned to Muncie, Indiana, in 1976 to
make a study of the city which could be compared with the earlier
Lynd studies reported:

> On [Tocqueville's] evidence and that of his contemporaries, the high point
> of social equality in the United States was probably achieved in the 1820s.
> The manuscript censuses of Middletown from 1850 to 1880 show us the
> outlines of a society that still looked very much like Tocqueville's America.
> Nearly every family had some property; none were very rich. Literacy was
> almost universal and advanced education almost unknown . . .
>
> And once the War between the States had settled the question of slav-
> ery, the few black families of Middletown lived side by side with their
> white neighbors, following the same occupations, sending their children
> to the same schools, working and voting with everyone else.[14]

Recent studies have challenged this idea that America was in this
sense a "classless" society. In particular, Edward Pessen has shown that
there did exist, especially in the cities, people of great wealth. According to Pessen, in about 1845 there were 113 millionaires in New York
City at a time when a million dollars was a huge sum of money. He
says, "By the 1830s hundreds of families in the nation's northeastern
cities had amassed great fortunes . . . By almost any criterion, opulent
Americans lived lives comparable to those enjoyed by their English and
Continental counterparts."[15] Indeed, the richest 10 percent in New York,
New Jersey, and Pennsylvania held some 40 percent of those states'
wealth. Pessen, thus, quite rightly insists that there was a considerable
inequality of wealth in the United States going back into the 18th century. Nonetheless, these families of huge wealth constituted only a tiny
fraction of the American population; the great central mass of the people belonged to a single class in which differences in wealth from top
to bottom were not large.

The division of American society into a middle and a working class
began after 1830, and was a direct consequence of the industrialization

of the nation. It resulted from the fact that large industries required a body of employees who worked with their heads instead of their hands—record-keepers, salesmen, accountants, planners, packagers, managers, and "clerks" of various sorts.

Added to them was another group of clerks who worked behind the counters in the new retail stores selling a variety of goods that were replacing the small specialty shops of an earlier time.

But it was mainly the white-collar crews in the mills and the factories which constituted the heart of the burgeoning middle class. As C. Wright Mills has pointed out, capitalism did not develop as a system of small enterprises, but centered on large-scale industries which could drive thousands of miles of railroad tracks through the prairies, and gouge thousands of tons of iron ore from mountain ranges. "The industrialization of America," Mills says, "especially after the Civil War, gave rise not to a broad spectrum of small businessmen, but to captains of industry." [16]

These new white-collar workers needed a different education and different attainments from those of the people on the shop floor. They needed to speak standard American English fairly well, to be able to handle numbers, to dress in a certain way. They were expected to have a measure of sophistication, a smattering of "culture," and a certain ease of manner. It was expected that these people would be paid more than the men and women on the shop floor; but more important than the pay differential was differences in attitudes between the new social classes. Perhaps most significant of these was the sense in the middle class that there was for them an opportunity to rise which did not exist for the laborers in the mills and the mines. In fact, there was probably less room for upward movement than there appeared,[17] but there was enough to give everybody hope; and it is probably true that the majority of the white-collar class did rise to one degree or another. (It should be noted that at first, and for some time after, these white-collar workers were mainly men, although women were coming to dominate in the schools, and eventually in the lower level clerical jobs.)

Very quickly the white-collar worker began to see himself as different from the worker at the machines. He was better educated, better read, richer, better dressed, more sophisticated. He appeared to himself as a substantially different sort of creature from the laborer. He was not alone in this view, for the laborer too saw him as something different. The working man may not have agreed that the white-collar worker was one of "the better sort," as the clerk might have put it; but that there was a difference he had no doubt.

Thus, by mid-century, there existed in the United States the beginnings of two distinct classes—white collar/blue collar, or middle class/ working class, whichever set of terms you choose to use. Sociologists today, it is true, have worked out a number of schemes for analyzing

America into classes, some of them using many fine divisions, others only a few rough ones. For example, "old money" with famous names sometimes see themselves, and are seen, as in a class of their own; and certainly the disfranchised at the bottom of the people—the addicted, people on welfare, and the like—are viewed by nearly everybody else as a class apart. But most modern sociologists would agree that the primary division in American culture—and indeed elsewhere in the industrial world—is the one that separates the white-collar middle class from blue-collar workers.[18]

The critical fact about the new middle class was that, for every so many laborers who were added to the workforce, an additional number of record keepers, salesmen, and planners was required. One clerk could make up only so many wage packets, type and file only so many letters, check only so many bills. The flood of workers coming from Italy, Russia, Hungary, and the American farmlands automatically swelled the middle class. And it was obvious that the bulk of this new middle class would have to be drawn from the old stock: the immigrants, in the period we are speaking of, were simply out of the race. Few of them could speak standard English, many of them were illiterate even in their own languages, most of them did not understand the folkways and customs that they would be required to follow in corporate offices, and many of them, as we have seen, were more concerned with collecting a stake to go home with, or worrying about their families, than with the upward mobility that was part of the psychology of the new middle class. Some of the children and grandchildren of the older German and Irish immigrants of the pre-Civil War wave were able to cross the line. But in the main the burgeoning new middle class was being filled with people whose names were Wilson, Cummings, Crane, Brooks, rather than Grossman, DeAngelis, Sing. This would in time change, of course. But the first settlers in the new American middle class had mainly British or Irish names, as indeed had the first settlers of the country. One study of Boston in the latter decades of the century showed twice as many of native-born as of immigrant stock in the middle class, even though immigrants were the majority in the city.[19]

It was, really, an explosion. It has been calculated that between 1870 and 1910, when the population of the country increased two and a half times, the new middle class jumped eight-fold. Of the total, only about a third were small businessmen, shopkeepers, and their clerks;[20] the rest were soldiers in the corporations.

It is critical for us to see that it was especially this native-stock middle class who took up Victorianism as their central ideal—indeed, belief system. These people were the classic American Victorians, the young men and women who fought their sexual impulses, campaigned against liquor, read elevating books and listened to elevating music, and raised their children to the high standard they held for themselves.

For these people the Victorian mode was more than just a moral idea; it was, in addition, a status code. Anyone who breeched the Victorian system was instantly suspect of coming from the immigrant lower orders. A person so branded would not only find his career endangered, but might discover that he was unwelcome in the homes of others of the middle class, and unable to find a marriage partner there.

The Victorian code, as it was refined by the new middle class after, say, mid-century, did not merely contain strictures against drunkenness and debauch: it was also composed of a complex set of manners touching on forms of speech, styles of eating, dress, and the like, down to tiny details as to how butter should be cut from the stick (sliced at right angles, not hacked off on the oblique), sleeves of a suit jacket creased (they weren't—sleeves were rolled), and a parent addressed (Mother and Father, never Ma and Pa.)

One small, but indicative dictum was that gentlefolk were supposed to know the maiden names of their four great-grandmothers. It was assumed that only people whose families had been in America for several generations could perform this feat. The point of much of this later Victorian code, then, was to separate the middle class from the uncouth barbarians swarming in from Italy, China, Poland. The code was a set of passwords: working people chewed gum, so middle-class children were forbidden to; working people drank, so middle-class people served lemonade; working people were noisy and gesticulated, so middle-class youngsters were taught to modulate their voices, keep their arms by their sides, never point.

As ever, we cannot carry this too far: it was not all snobbery, for many, and perhaps the majority, of the Victorian middle class did genuinely believe in temperance, sexual control, and decent behavior as being of benefit to the society as a whole. But as Victorianism came more and more to be seen as a class designation, it behooved people to cling to it ever harder. And by the later decades of the 19th century, the middle class and Victorianism were virtually synonymous.

The creation of a distant proletariat and middle class was the most significant social consequence of the new industrial society. By 1920, according to the Lynds, the division was rigidly established. "It is after all this division into working class and business class that constitutes the outstanding cleavage in Middletown. The mere fact of being born upon one or the other side of the watershed roughly formed by these two groups is the most significant single cultural factor tending to influence what one does all day long throughout one's life; whom one marries; when one gets up in the morning . . . and so on indefinitely throughout the daily comings and goings of a Middletown man, woman, or child."[21]

Before some moment in the mid-19th century, it was possible to speak of the central mass of Americans—all those millions of farm families—

as if they constituted a single group, sharing many habits and beliefs. By 1870 or so, we cannot so easily generalize, but must specify which class we are talking about. In Middletown of the 1920s the ratio between the working class and what the Lynds called the "business" class was 71 to 29; the middle class then constituted about 30 percent of this small American city.[22] Today the balance would be closer to fifty-fifty, and the manners and mores of the two groups would not be so distinctly marked.[23] But the division, if unspoken, and perhaps unspeakable in many circumstances, nonetheless remains.

For the middle class, in the early years of the new industrialism, the city was a very comfortable place. These people could afford apartments and houses that by today's standards were large. Most of them could afford help of one kind or another, and it was hardly uncommon for middle-class families to have a live-in maid; Irish women in particular gravitated to domestic work, and "our Bridget" was a common figure in middle-class homes.[24]

The wives of the middle class did not work. Despite the growing strength of what was then called "the woman movement," the paternalism of the Victorian ethic still was powerful, and it required the women to devote themselves to raising the children and maintaining a home free from evil influences—a home that would act to encourage moral and spiritual uplift.

This middle class, although growing, in the decades on either side of 1900 constituted no more than a quarter of the population of the United States. Nonetheless, it was the dominant section of the social system. It staffed the executive offices of the burgeoning industrial machine, it supplied the majority of office-holders in national, state, and to a lesser extent city governments, it created the art and literature of the time, and perhaps most important, it set the style which those who hoped to rise must follow. In a certain sense this Victorian middle class would— for the moment—decide what America was all about.

The prosperity of this new middle class was bought for them by the working people, and especially the immigrants. Of course some of the immigrants did manage to work their way out of poverty and establish themselves in the society. But the great mass of working people, in the years of the massive immigration, lived in a poverty that makes the poverty of today look like affluence. One contemporary report said that only 10 percent of tenement families met a good standard of family life,[25] and this was no very high standard they were being judged by. Another authority estimated that 10 million slum dwellers did not earn enough to give them a reasonable standard of living,[26] and again this was a standard which we would not consider reasonable today.

Many of the tenements lacked running water. Toilets were in some cases outhouses in backyards six flights down for dwellers on top floors; and even when there was running water, the toilets were in the halls, shared by everybody on the floor. (Such conditions still existed as late as the 1950s.) Few tenements had central heating: warmth was supplied by small coal stoves. As a rule slum families were crowded into one or two rooms, with perhaps four or five people sleeping in a room, or even in a single bed. Families with larger quarters generally took in boarders in order to help pay the rent; given the surplus of immigrant males, there were large numbers of single males who needed places to sleep.

Diets were poor. There was never really enough money for anything, and the working people of the slums ate the cheapest kind of food. "Urban immigrants . . . rarely had a balanced diet, often scarcely any food at all . . . Their diet was frequently heavy with starches and fats and low on protein."[27] As a consequence, everybody had to work; children as young as ten might go into the sweatshops, or do piecework at home. By fourteen a youth was expected to do an adult's work. Most people worked ten or twelve hours a day, six days a week, but people doing piecework at home might work into the night until sheer exhaustion felled them, in order to make the day's pay match the day's expenses. Single mothers, who in that time were more likely to have been widowed or abandoned by their husbands, rather than unwed, suffered in particular from this kind of exploitation.

All of this was fearfully destructive of humankind: William Dean Howells, perhaps the most famous American literary man of the time, described "the work-worn look of mothers, the squalor of babies, the haggish ugliness of old women, and the slovenly frowziness of the young girls . . ."[28] There can be no doubting the truth of this statement: the photographs made by the reformer Jacob Riis, which he published in a scathing report on the immigrant slums titled *How the Other Half Lives,* showed a level of human degradation beyond anything we would believe today: men head down on tables in all-night two-cent restaurants, people sleeping five and ten to a tiny room for "five cents a spot," homeless "street Arabs" of ten and twelve huddled together in doorways or on sidewalk gratings.[29]

Again, it must be borne in mind that not all immigrants sank to this level. Tens of thousands of them enjoyed their Saturday nights at the saloons, their Sunday afternoons at the beer gardens, a trip to the baseball park, a picnic or summer's evening boat ride sponsored by the union. But huge numbers of immigrants were living in rock bottom poverty. In the Pennsylvania mill town of Homestead, half of all black and Slavic families were living in one- and two-room shanties, without toilets or running water. It is no wonder that an incredible two-thirds of their children died before they reached the age of two.[30]

The United States, thus, by the 1890s had become a country divided into the haves and the have nots. But the new industrial-urban culture had what may have been even profounder and certainly subtler effects on the lives of Americans than dividing them into a nation of the poor and the prosperous. The movement from farm to factory, from country to city, worked dramatic changes in the ways people related to the world around them—for, of course, it was a new and different world they were relating to.

For one thing, the tightly enclosed social group of the more or less self-sufficient farm family, living within a relatively small social group, was gone. No longer did people eat and work and worship and play with the same group of family and friends. The social group was now splintered. Each morning families would separate, the children going off to school, the adult and sub-adults to various jobs. The evenings, too, were often spent separately, with the young people out on the streets in search of amusement, the husband at his club or the local saloon, the wife at home doing household chores. The same pattern obtained on weekends, when the parents might go to the beer garden, while the younger people courted or sought amusement elsewhere. People began to find that they belonged to not one, but to several social groups—the family group at home, which might have included boarders and other non-family members; the group at work; and the recreational group in the saloons or the street corner and park hang-outs where youths met in search of excitement and in many cases, sex.

There were some countervailing forces. Immigrant working-class families were held together by the facts of their economic lives. John Bodnar has pointed out that new immigrants counted on kin already established in America to find them jobs, places to live, social groups.[31] The consequence was that groups of relatives frequently worked together in the same mill, mine, construction site. Moreover, because in many, if not the majority of, instances a single income could not support a family at even subsistence level, wives, children, and other kin in the household were under heavy pressure to contribute to the family income at early ages, and to go on doing so as long as it was necessary.[32] "Family members were continually instructed in the necessity of sharing and notions of reciprocity were constantly reinforced."[33]

The middle class, too, attempted to enforce conformity to family rule, especially in order to make children hew to the Victorian line. With their bigger homes, the omnipresent piano, separate rooms where people could gather to talk, sing, play Parcheesi or Go Fish, there was simply more to do at home. Nonetheless, the primary effect of the city on the family group was centrifugal. By the time of the *Middletown* study in the early 1920s, youths even of the middle class were spending substantial time away from their homes. Less than a fifth of *Middletown* teenagers spent half their nights at home, and the Lynds noticed no difference between working-class and business-class kids in this respect.[34]

Even more significant to the splintering of the social group than the daily dispersal of the family was the increasing number of single people flooding into the cities in the last decades of the 1800s. To begin with, in almost every group—the notable exception was the Irish—male immigrants outnumbered females, sometimes by substantial percentages. Thomas J. Archdeacon, speaking of the "new" immigration of the later decades of the century, says: "A larger proportion of the immigrants was single, young adults than previously; a greater percentage of them was male; and more of them saw the trip across the ocean as a sojourn rather than a permanent relocation."[35] Fifty-four percent of Jewish immigrants were male, three out of four Italians,[36] 90 percent of Chinese, Balkans, Turks.[37] Huge numbers of immigrants, then, were young, single people, and the male surplus among them eventually ran into the millions.

Conversely, in the old stock leaving the farms for the cities the females outnumbered the males. The reason was obvious: given the new industrial economy there was less for women to do on the farms. As a consequence, as early as 1880 there were 7 percent more males than females on farms.[38] In 1900, females outnumbered males in the big cities by a few percentage points although, as the immigrants continued to flood in through the early decades of the 1900s, males very quickly became more numerous than females.

These young single people lived mainly as casual boarders in crowded tenement homes if they were working class; in more genteel lodging houses, where they would have a room to themselves, if they were white-collar workers. Many of them were in their mid-teens—fourteen, fifteen, sixteen. They were unsupervised, and largely unsupervisable. They could do as they liked, spend their money, however little it was, as they liked, mingle with whomever they chose. Such an arrangement, with numbers of young single people afloat in the social system, was a startling novelty in human life. Heretofore almost everywhere it was customary for the young to stay under the parental roof until they formed families of their own. Indeed, in 17th-century America it was actually forbidden for people to live alone, and single people, like orphaned children and new immigrants, were placed in families in order that "disorders may be prevented and ill weeds nipt."[39] By the 18th century living alone was legal, but was considered eccentric, perverse, or even a sign of madness. Says one authority, "single adults of any age living alone were very unusual, and lifelong bachelors and spinsters were a rarity."[40]

Between everything, then, in the new industrial world relationships were more varied, but less intense—more "diffused."

This idea that the modern city was responsible for what has been called "alienation" is hardly new: the German philosopher and sociologist Georg Simmel, who died in 1918, was already taking note of a "reserve with its overtone of hidden aversion" which he descried as a "general mental phenomenon of the metropolis."[41]

To measure anything so intangible as alienation is difficult, but common experience tells us that the sudden disappearance of a work-mate, or a drinking companion, is not nearly so earthshaking as the departure of a child, parent, ancient friend, admired minister. There were many associations in the lives of people in the industrial city—co-workers, saloon or streetcorner companions, and landladies and fellow boarders who came and went, familiar shopkeepers, and, in many instances, series of sexual partners of a kind that were rare in the countryside. But relationships with these fellow city-dwellers were not tightly bonded. In the new society a substantial proportion of the people you spent time with were only lightly linked to you. Relationships were less emotionally charged.

Adding to this diffusion of relationship was the fact that, especially in the working class, people changed jobs frequently. In Middletown, half of workers typically held their jobs for five years or more, but nearly a third changed jobs every two years or less.[42] Job changes frequently meant changes of residence; and people, especially in the immigrant slums, very often moved every year in hopes of finding a better place, because they were behind in the rent, or for other reasons. Bodnar says that immigrants "were constantly on the move as were most residents of urban-industrial America in the nineteenth and early twentieth centuries."[43]

The consequence of both job and residential mobility was again to loosen personal ties. A Jewish or Italian immigrant might live in an ethnic ghetto which contained something of the atmosphere and folkways of his old home, but there was no real building up of long-term ties to a community and a group of people. In Middletown the Lynds noticed of working-class husbands and wives that, "many of them have few if any close friends."[44]

A second difference in human relations between the farm and the urban system was that in the new world *specific* people were less necessary to each other in a basic way. Of course the industrial city was a tightly meshed system in which it was necessary that millions of people correctly perform the functions they were assigned. But it was all abstract: people were to a considerable extent interchangeable parts.

In the older agricultural system people needed *specific* other people in real ways. Only family members would perform certain essential functions for each other. Once they were gone, they were difficult to replace. There was a very high degree of mutual dependency in farm families and communities, and these dependencies forced people to live inside of intense personal bonds.

In the industrial city no single person was indispensable; nobody was really very important to the operation of the system as a whole. The community was simply too large for anybody to identify with. The sense of belonging to some surrounding social group was diminished, if not done away with altogether. In the new industrial society people might take a certain pride in being New Yorkers or New Orleanians; they might feel a certain loyalty to a church society, an ethnic group, a service club, an athletic team; and of course on the management level, at least, employees were expected to act on a sense of obligation to the corporation. But the bonds of membership were weak: people moved from Chicago to San Francisco, changed jobs, quit clubs, without much sense that they were letting the team down.

A third major change in the way people thought and felt with the rise of the industrial city had to do with people's vision of the future. The farmer was intensely future-oriented. When he planted in the spring he was not merely thinking about the harvest in the fall; he was, instead, calculating what crops he would need until the succeeding harvest some fifteen months away. When he planted an orchard he was planning five or ten years ahead. And when as his children matured he decided to cultivate additional acreage in order to provide farms for them when they married, he was thinking in terms of a generation.

In the new industrial city the majority of people did not have futures—or rather, they had futures which would not be materially different from their presents, and might well be worse. The immigrants, Bodnar says, "expressed hopes for a better future, but were usually too busy making ends meet to do much about it."[45] It was a dream, not a course of action. To be sure, most of the 25 percent in the growing middle class were ambitious, upwardly mobile, and saved money for bigger homes and worked for promotions. Some of the native stock coming off the farms had plans to get ahead; some of the immigrants were saving a stock of capital to take home, and others were making their ways as small entrepreneurs. But the majority of the citizens of the big industrial cities were not going anywhere, except sideways. For immigrants, "Opportunity was not vertical but horizontal, a fact which tended to blunt any rhetoric of social mobility immediately upon arrival."[46] Running like a dark thread through the *Middletown* study is the sense of discouragement, bordering on despair, felt by working men and women. For many of them sporadic patches of unemployment were the norm; most of them saw that once the men reached their forties they would become increasingly unemployable; few of them really expected to rise.[47] The working man, the Lynds concluded, lived "in a twilight zone of doubt as to whether he can actually get ahead."[48]

For the majority of Americans in the new industrial city there was no real future, only an endless present. It was almost impossible for them to save anything; and even when they could, after ten or twelve hours in a mine or a sweatshop the temptation to spend a little for a good time was almost inhumanly hard to resist. Perforce, these working people, and many white-collar people as well, tended to think in the short term, often the very short term, simply counting the hours until the closing whistle blew, counting the days until they were handed the next pay envelope on Saturday. And if they should spend, after paying the landlord, whatever they had left in that envelope on beer and dance halls, who could blame them?

Finally, the new industrial society produced a sharp division between work and play. Back on the farm there was as a rule no chunk of time set aside for amusement. Things ran according to natural rhythms. People worked long and intensely during planting and harvesting seasons; given a spell of good hot weather for haying, they might work in the fields by moonlight to beat an impending rain. In winter the pace slowed, especially in northern areas; people made day-long, even week-long, visits to friends and relatives. Pleasure was woven into the fabric of life where it fit. Jack Larkin has noted that these earlier Americans had far fewer official holidays than was the custom in the Old World.[49] The work ethic prevailed, but enjoyment was taken where it could be. Women might swap gossip or tell stories as they sat before their spinning machines. Corn was often shucked at competitive husking bees, which might be accompanied by singing, and followed by dancing and a communal meal. Barn-raisings were usually an excuse for a frolic, with food, drink, wrestling matches. Community dinners were part of the routine of Sunday worship.[50] The life of the farmer was hard and often extremely busy; but there was room in it for people to stop for a moment to drink a cup of tea, to gossip, to pull from the cider jug sitting in the shade of the big maple in the middle of the hay field.

The new industrial society changed all of that. Now work began and ended on an instant, with the blowing of a whistle or the ringing of a bell. Although some workers, like the women in the sweatshops of the garment trades, frequently were able to gossip, it was not possible in many work circumstances. Construction gangs were not encouraged to get up wrestling matches during work hours. As a consequence recreation became something it had never quite been before, for the mass of people at least—that is, an activity of its own, rather than as an adjunct to a task, a celebration.

The fact that the work day, however long and arduous, ended at a stroke, meant that for many people there was time to fill. This was hardly true of everybody: mothers trying to raise children on pittances were likely to be busy for much of their waking hours, and in many families people worked at second jobs or did piecework at home in the evenings. For some, work was almost without end. But there were many others, especially the millions of the young and single who lived in the industrial cities, who found themselves presented each day with a block of time to do with as they pleased. This was certainly true of the middle class, whose hours were shorter than those worked by the people on the factory floor, and whose children increasingly through the early days of the twentieth century had fewer chores, held fewer after-school jobs. But it was also true of huge numbers of working people.

These city-dwellers could have done many things with this spare time. But in fact, more and more, they chose to amuse themselves—go to a dance hall, see a show, visit a saloon, look for a sexual adventure. And on this spare time, as we shall see in more detail soon, was built the huge entertainment industry which so dominates American culture today.

All of these new attitudes, habits of mind, ways of doing things brought forward by the rise of the industrial city pressed in the same direction. The loosening of family bonds, the disappearance of the community in the old sense, the tendency to live in the present, the increasingly transitory nature of relationships occasioned by frequent moves from one neighborhood to the next and the equally frequent changes of jobs, worked to separate human beings. No longer sharing common purposes, needs, rewards, and duties, they became more and more self-contained, isolated units rushing past each other without acknowledging each other, without mutual interaction, looking mainly to serve their own interests. The American city, says Howard P. Chudacoff, "has always been a divided and divisive place."[51]

I am not suggesting that the family suddenly disappeared, or that people did not care for each other anymore. There were still, in the early 1900s, millions of people on family farms, still millions of people living in the close communities of small towns. For the middle class, the family—however much its members went their own ways—was the central institution; and, as we have seen, to millions of immigrants family cooperation was essential to survival.

Nonetheless, the family was becoming more loosely knit, and less tightly meshed to a community; indeed, the community, in the old sense, was disappearing. As a consequence, the basic unit of society was becoming the self. And thus the seed was planted for the growth of a new ethic in which the first loyalty was not to the community but to the self; in which the first duty was not to the family but to the self. It was a new idea; never, so far as I have been able to discover, has there existed a culture in which the needs of the self always—or nearly always—are seen as coming first, ahead of the needs of the family, the community, the nation, humankind in general. But in the new industrial city it only seemed to make sense. And it followed ineluctably that if the self was the significant social unit, people had a right, even an obligation, to gratify themselves.

CHAPTER 5

The Institutionalization
of Vice

The industrial city gave rise to a plethora of institutions that today we take for granted: ward politics, a vast entertainment industry, a huge system of libraries and museums built for the enlightenment of the public, and much else. A most interesting—and in its way characteristic of the late Victorian period—was what was called at the time the segregated vice district.

Its impetus was the terror of venereal disease, especially syphilis, which haunted the Victorian Age. The disease could maim; it could kill. Every year it crippled thousands of babies in the womb. The final stage of the disease, tertiary syphilis, which might not become evident for many years, could bring blindness, insanity, a lingering death. To some extent the terror of syphilis was a product of the Victorian imagination, already "half in love with easeful death," as John Keats, an early exponent of melancholy, wrote. But syphilis was sufficiently real to engender terror. It was almost as deadly as our present scourge, AIDS, and it cast a vastly wider net. According to one source in the late 19th century, New York City was reporting 243,000 cases a year; Pittsburgh, 20,000;[1] and Minnesota authorities estimated that between 5 percent and 18 percent of the population was infected, depending on the county.[2]

Venereal disease was a most serious and frightening fact of Victorian life, a shadow hanging over everything. It was a subject nice people did not talk about, nor read about in their magazines very much. But there was an acute, covert awareness of it. Every doctor again and again saw children born with congenital syphilis of innocent wives who had been infected by husbands who had themselves perhaps picked up the disease in the course of a few youthful visits to brothels. It was a common tragedy, and the disease was seen by public health authorities as one of

49

the leading problems confronting the social system of the entire western world.

It was widely agreed that a prime source of infection was the prostitute. A study of prostitutes in Baltimore reported that of 289 women examined, 63 percent had syphilis, 92 percent had gonorrhea, and only 3.4 percent were disease-free.[3] While this study may be open to question, it is reasonable to believe that a prostitute having sexual intercourse with upwards of a hundred men a week is more likely to be infected than the general run of the population.

During the Civil War, Union army officials, troubled by the spread of venereal disease through the troops, turned, at certain times and places, to a system of "regulated" prostitution[4]—the common term. This system required that prostitutes be inspected at intervals; those free of the disease would be permitted to practice their trade, those infected would be hospitalized. This system had been in place in France for some time, and was, in the 1860s, being tried in England.[5] Its major weakness, of course, was that a prostitute could easily contract syphilis an hour after being cleared by medical examiners, but it was nonetheless considered to be valuable in keeping down the disease. It began to seem to municipal officials in the growing cities and towns across the United States that it was the best—perhaps the only way—to diminish a major public health problem. The idea of regulated prostitution got enthusiastic support from some quarters, especially the medical profession, police forces, and municipal authorities, and in July 1870 the city of St. Louis put into law a system of regulated prostitution. Between then and March of 1873, 2,685 women were registered—about 1.4 percent of the entire female population of the city.[6]

However, there was substantial opposition to regulated prostitution, coming especially from women's groups, who objected to the double standard by which men were free to "sow their wild oats," as the euphemism went, while women were supposed to remain chaste before marriage. The solution to the venereal disease problem, so this line of reasoning went, was to impose the same standards on men and confine sex to the marriage bed. Prostitution should not be encouraged through regulation but stamped out.[7]

The anti-regulation forces prevailed, and St. Louis was forced to withdraw the laws. Efforts in other cities to introduce regulated prostitution were also defeated, usually by a combination of women's groups, the clergy, and middle-class subscribers to the Victorian ethic in general. But the de facto victory went to the regulators: the police and municipal authorities. They were charged with enforcing the laws, not the reformers, and when they chose not to enforce them, there was little the reformers could do. (It was this fact, as much as any, that gave force to the drive for the women's vote: it was believed that women would universally vote for reformers who would then take over municipal governments.)

Municipal authorities and the police were everywhere generally strongly in favor of regulation. Their motives, to put it charitably, were mixed. Few were actually indifferent to the spread of syphilis, a highly democratic disease that struck high and low together, and most did genuinely believe that regulation was the way to deal with the scourge. But police and municipal officials had other reasons for favoring regulation. For one thing, when left to their own devices, prostitutes tended to parade up and down the more fashionable streets, where they were likely to encounter wealthier prey. This meant that the young from the better families were being exposed to temptation. It further meant that property values in these streets would fall. The city fathers of St. Louis were quite explicit about this. They said: "The most fashionable thoroughfares were thus converted into public brothels, where many were allured to sin and disease by the fascination of a moment, who did not and would not have designedly placed themselves in the way of temptation. The concentration of prostitutes [in a regulated area] is also desirable for other reasons. By the judicious selection of suitable situations for houses of ill fame, the value of surrounding property is depreciated in the least degree, thereby benefiting private interest and the public revenue."[8]

For yet another thing, as the unofficial movement toward segregated red light districts grew, it became clear to many people that there was a lot of money to be made, if one were so inclined. As the trade was officially illegal, it allowed police officials all the way up the line from the cop on the beat to the commissioner himself to extract from the business steady and quite handsome payoffs. This system of bribes was open and blatant. One New York City reformer, Charles Henry Parkhurst, in an 1895 report on what was frequently known in middle-class circles as the "social evil," said that one brothel keeper was paying a police captain $500 a month[9] at a time when an ordinary laborer would be lucky to make $50 a month. This figure is substantially higher than others given elsewhere; but even a small brothel might well pay $50 a month, plus $5 or $10 per girl—a term I use advisedly—for a total of perhaps $100 a month,[10] which would be distributed among a number of people. A police officer with a salary of a few hundred dollars a year might easily double it through his share of the payoffs; police lieutenants, captains, and on up could make correspondingly more. In a major study of the Chicago red light districts, Walter C. Reckless said: "Affiliation, almost inevitable in American city politics, between vice and politics, was a heritage from an earlier period, when the policy of the police was to keep up a decent appearance, while tolerating vice. This had assumed the character of a system."[11]

The police were not the only ones making money. Because public, if unofficial, policy was to keep the vice districts out of middle-class neighborhoods, they were almost always located in poor areas. Landlords could charge a brothel keeper grossing several thousand dollars a month far more than they could immigrant families living on marginal incomes. Renting to madams was exceedingly profitable, and a great many respectable people with real estate holdings in slum areas owed their prosperity to the vice trade.[12]

So lucrative was renting to the vice trade that in the first decade or so of the century numbers of investors were building houses specifically designed as brothels.[13] In Washington, D.C., for example, "One real estate syndicate is known to have built a row of houses especially for use in enterprises of prostitution. A block of large 'parlor' houses, modern in construction, and each with a private dining hall, were built in a secluded section of the city just five squares south of the United States Capitol . . . Although the location of the 'new Division' was supposed to have the sanction of the authorities, the strictest secrecy was maintained as to the uses to which the houses were to be put."[14]

Thus, a number of groups in the social system stood to benefit financially from the creation of the segregated vice district; it was simply much easier to extract high rents, bribes, and the rest of it from an organized system of prostitution than from free-lance whores operating out of rooming houses and cheap hotels. The reformers may have had justice on their side; but the promoters had the money and the police power, and they let the business grow.

Thus, in the years from the end of the Civil War until the reformers finally triumphed after the turn of the century, there came into existence in every major American city, most small cities, scores of small towns, and even some villages, segregated vice districts. In the biggest cities there were usually two or more such areas: New York City had, besides the famous Tenderloin in the West 30s and 40s, vice districts at various times in Chinatown, on the Bowery, and later in Harlem;[15] in Chicago the famous district was the Levee in the First Ward (the boundaries shifted over time), but there were other districts as well.[16] In San Francisco there was the famous Barbary Coast, a major tourist attraction for the city; in Washington it was the Division; in Newark, New Jersey, the Coast; in Atlantic City, the Line. There were vice districts everywhere: in Cheyenne, Wyoming; Fall River, Massachusetts; Lincoln, Nebraska; Mansfield, Ohio; New Haven, Connecticut; Peoria, Illinois; in Louisville, Detroit, Denver, Minneapolis, Milwaukee, Seattle, Omaha, Rochester, Salt Lake City, San Diego, Kansas City, Syracuse, Cleveland, Pittsburgh, Boston, Philadelphia.[17]

The segregated vice district was a product of the industrial city. The brothel was hardly an invention of the nineteenth century: the brothels of Pompeii at the beginning of the Christian era were famous. In the

United States in the eighteenth century Philadelphia had a vice district called Hell Town,[18] and by 1820 New Orleans had one called the Swamp.[19] But as a rule in the early days brothels and street walkers were scattered, and "Outside of a few cities, prostitution had not really become widespread."[20] D'Emilio and Freedman report that in 1720 the Virginian William Byrd, who kept a very frank diary, could not find a prostitute in Williamsburg, then one of the most considerable towns in Virginia.[21] The point, quite simply, is that most people lived in places that were too small to support prostitution, to conceal it, or to permit it.

The new industrial city produced conditions conducive to the trade: a large deposit of single males, the necessary anonymity, a loosely constructed social system which did not inculcate a strong moral code and could not enforce its sanctions in any case, and a supply of women—girls, really—who found advantages in the brothel life.

We are fortunate that, in the early years of the twentieth century, a reformist movement, coupled with a new interest in the social sciences, produced a large covey of reports on the local vice districts, usually instigated by an official, or at least semi-official, vice commission. Such reports were produced in Pittsburgh, Philadelphia, Minneapolis, Louisville, Little Rock, Honolulu, Elmira, New York, and elsewhere. At least thirty were issued between 1910 and 1915.[22] Some reports were long and extremely detailed, and taken together they give us an astonishingly clear look at this important American phenomenon.

Vice districts were, with one exception, illegal. The exception was New Orleans' Storyville, which had a legal existence from January 1, 1898, to the fall of 1917, when it was closed by the Navy Department in a high-handed and certainly illegal maneuver which was protested by the city fathers.[23] But most districts operated as if they were legal, paying a fixed schedule of "fines" which were in fact operating fees. For example, in Minneapolis in the 1880s the police installed an indirect licensing system in which each house was charged $50 a month, each woman $5 or $10 at different times. In 1897 reformers pushed through a law raising the "fines" to $100 a month, with the hope that this would be more than the madams would be able to pay. The police countered by collecting the fines only every other month.[24] In Little Rock the fees were $25 a month per house, $5 a month for each girl. Authorities charged another $50 a month for the right to sell beer, a commodity which probably produced greater profits than the sale of sex. These fines were "treated by [the madams] as a license for running their business," the Little Rock vice report said.[25]

Some vice districts were nationally famous, especially Chicago's Levee, New York's Tenderloin, New Orleans' Storyville, with San Francisco's Barbary Coast certainly the most celebrated of all. These districts—to use the generic term commonly employed—have been invested by recent writers with a good deal of glamor, and even at the time they were seen by some people not familiar with the actuality as filled with beautiful young women in elegant clothes, lounging around fancy parlor houses decorated in plush, ormolu, cut glass, with champagne and sprightly music.

And indeed a few such places did exist. Perhaps the most famous of the brothels in the heyday of the vice districts was Chicago's Everleigh Club. (Everleigh was said to be the last name of the sisters who owned it, but it is suspiciously close to a sexual pun.) According to one writer, in any case, the Everleighs were "upper-class Kentucky girls."[26] Ada was twenty-three and Minna twenty-one when they opened their house February 1, 1900. They charged $50 a visit, an enormous sum for the time. "Girls were given daily instructions in etiquette and told to use the library which had been installed. 'Stay respectable by all means; I want you girls to be proud you are in the Everleigh Club.'"[27] The club had "theme rooms" with Moorish, Japanese, and Egyptian decorations, and "a famous Gold Room which was refinished each year in gold leaf and polished industriously every day."[28] Instead of the banjos and honky-tonk pianos most houses favored, the Everleigh Club featured string ensembles and gave an annual Christmas party for the press with champagne, dancing, and gifts. So famous was the place that when Prince Henry of Prussia visited Chicago he asked to see it. At the special party arranged for him the girls were dressed in fawn skins and performed dances intended to suggest a Roman orgy.[29] When the Everleigh sisters were finally put out of business by the reform movement in the fall of 1911, they were reputedly worth over a million dollars.

Today the best-remembered of the glamorous brothels is certainly Lulu White's Mahogany Hall in New Orleans' Storyville. Mahogany Hall became a legend in the mythos of jazz, although very little jazz was played there; it has figured in books and movies about New Orleans jazz, and its name was immortalized in the tune "Mahogany Hall Stomp," written by Spencer Williams, Lulu's nephew, who lived in the brothel for a number of years, according to Al Rose.[30] Lulu White's, according to its promotional brochure, cost $40,000 to build, was made of marble, and was four stories high. It had an elevator and each room had a private bath. Lulu White was an octoroon and her house featured light-skinned women of mixed blood. Jelly Roll Morton said that the mirror parlor in the place cost $30,000, almost as much as the building itself, and that there were mirrors at the head and foot of each bed.[31]

But the Everleigh Club and Mahogany Hall represented tiny islands of glamor in a vast sea of sordid brothels which offered swift commercial sex undertaken in dreary rooms on beds sometimes covered with oilcloth, or even on tables and benches in the booths of saloons. The Hartford, Connecticut, vice commission gave this description:

We passed a saloon and a restaurant and then opened what appeared to be a stable door; up a flight of stairs into a low-ceilinged room; at one side was an opening into the wall not as high as a table, supposed to be a "get away" into the next house. A young man sat there, a typical tough. He was the "checker." The checker is the man who takes the money from the customer before the man can go into a room with a girl. The check is given by the customer to the girl and she is supposed to turn over the checks to the proprietor . . . There were two bedrooms divided by a wooden partition. The two beds had no linen on them, just a piece of dark, dirty, oilcloth; there was a pillow on each with a case on, which was the only piece of white on the bed. There were three girls there. They appeared to be not more than seventeen.[32]

This was a fifty-cent house, the bottom price generally, although Rose reports that some Storyville crib girls sold themselves during slow times for a dime.[33] (Again we must remember that in that time and place a nickel bought enough red beans and rice for a meal.) The girls in many of the cheaper houses usually wore kimonos, a reflection of the vogue for Orientalism of the time, which they had to buy at inflated prices out of their earnings. In the more common dollar houses there might be sheets on the beds, and the girls might be somewhat better dressed, but the atmosphere was not much more elevated. From this broad base of fifty-cent and one dollar brothels, the houses went up in quality pyramidally. Leaving aside the few elegant houses on the order of the Everleigh Club, the most expensive brothels had a top price of about $5. In Little Rock, for example, the prices ranged from $1 to $3 in the white district.[34] (In the South, of course, brothel districts were invariably segregated; even Storyville was divided into two distinct parts for blacks and whites.) The general rule was that the fee was split equally between the madam and the prostitute, but as the madams routinely overcharged the women for their food, laundry, kimonos, beer, and the like, the prostitutes were frequently in permanent debt to the brothels.[35]

Nor were the girls invariably great beauties. In the early part of the century a professional photographer named E. J. Bellocq, a hydrocephalic dwarf, took a series of pictures of the girls of Storyville, some of which have been published.[36] It shows a group of perfectly ordinary looking young women, many clearly teenagers, some quite pretty, some fairly plain, and the majority rather hefty, the taste of the time in part mandated by the fat-rich diet of the Victorians.

As the Bellocq photographs suggest, the prostitutes of the period were appallingly young. Virtually all of them were under twenty-five, with the majority teenagers. Something like a quarter of them—the figure is calculated from various discussions in the vice reports—became prostitutes at fifteen or under.[37] One survey said that 41 percent of the girls studied had become prostitutes at sixteen or younger.[38] In St. Louis, of the registered women seven out of eight were under twenty-one.[39] There are valid reports of girls being prostituted as young as nine years old,[40] and perhaps as many as 5 percent were prostitutes at thirteen or younger.[41] In sum, it is a fair estimate that at any given moment half or more of the "women" in the vice districts were girls of eighteen or younger. The "pretty babies" of legend were in fact that.

It was—and still is—widely believed that girls were driven into the trade primarily for economic reasons. There can be no doubt that the prevailing low salaries and long hours made brothel life look attractive. Sears, Roebuck was paying its female apprentices, who were mainly teenagers, from five to eight dollars a week. The Chicago Vice Commission reported that the average wage for girls working in department stores was six dollars a week, and went on to ask, "Is it any wonder that a tempted girl who receives only six dollars a week working with her hands, sells her body for twenty-five dollars per week, when she learns there is a demand for it, and men willing to pay the price?"[42]

But there was more to it than money. Most teenage girls did not become prostitutes. Vice commissions, which were naturally interested in what turned girls to prostitution, gathered a good deal of information on the subject. The girls themselves mentioned low wages as the motivation only infrequently. In the Hartford report, of 100 women questioned only six mentioned low wages.[43] There were other factors at work.

To begin with, a substantial number of these young women probably were of below-average intelligence. The Massachusetts Vice Commission reported that 51 percent of their sample were "feeble-minded" to the extent that they eventually had to be institutionalized. "Of the 135 women rated as normal, only a few ever read a newspaper or a book, or had any real knowledge of current events, or could converse intelligently upon any but the most trivial subjects."[44] Ruth Rosen, in her careful study of American prostitution, has argued with the contention that high percentages of the women were mentally deficient. She says the fact that "many prostitutes expressed contempt for middle-class niceties and values was offered as strong evidence of their feeble-mindedness."[45] It is certainly true that many middle-class reformers were stunned to discover that most of the prostitutes they tried to "save" did not want salvation if that meant going into the sweatshops at six dollars a week; and they often ascribed this unfortunate stubbornness to mental deficiency, or a character defect.

Nonetheless, a 1919 study of "mental deficiency" of prostitutes, using the then standard Binet-Simon test, concluded that 53 percent had a mental age of ten or under.[46] Other studies show a range of 30 percent to 97 percent of the prostitutes to be feeble-minded, with the bulk of these studies in the 35 percent to 71 percent range.[47] Many of these studies can be dismissed as incompetent, probably all of them were saturated by class and ethnic bias, and further distorted by the fact that so many prostitutes were functionally illiterate. Yet too many of them say the same thing to be dismissed entirely. It is probably safest to say that, as a class, these young women were below average in intelligence, but to what degree we are not sure.

Perhaps more important, huge percentages of these women came from poor family backgrounds. The Pittsburgh Vice Commission reported that almost half the women studied were without parents, and only 15 percent had both parents alive.[48] The Massachusetts report added that "nearly all come from families in adverse circumstances. Immorality, drunkenness and crime are usually a part of the early history."[49] The Pittsburgh Commission concurred: "Low wages are an indirect factor as they produce an unattractive life, parental neglect, a taste for cheap amusements and strong stimulants, and similar elements in the psychology of poverty."[50] None of this should surprise anyone: studies made decades later show that even today promiscuity in adolescent girls is associated with father-absence.

This, then, was the real problem. Many of these girls were the children of immigrant parents who were finding it difficult to make their ways in an alien society. There was never enough money for the necessities, much less the amenities, and in too many cases the fathers had disappeared. Whether born of immigrant parents or not, it was customary for working-class adolescents to become wage-earners at fourteen, if not before. Everybody had to work once they were old enough, and sometimes when they were not old enough.

Necessarily most of these girls turned over most or all of their wages to their parents. Typically they might be given an allowance of fifty cents a week, which might buy them an occasional frivolity or evening out to see a show, go dancing, or drinking.[51] Most of the time they were without money to buy themselves stockings or even small treats like candy.

Furthermore, there was nothing for them to do in the small, abysmally overcrowded tenement apartments they inhabited; and this overcrowding itself provided an education in sex. Said one contemporary report, "Hardly a married couple in any crowded neighborhood have a room to themselves and the children sleep with their parents up to the approach of youth. It is almost inevitable that children should come to know of the innermost reserves of marriage, as a result of which many are led to surrender their chastity and even to participate in gross immoralities."[52] Brothers and sisters frequently shared a bed beyond puberty, and there was always the ever-present lodger. The Philadelphia report said, "There is, among boys and girls of school age (from eight to fourteen), an appalling amount of experiment in sex familiarities, and even coition."[53] What is surprising is not that there was a lot of juvenile sex in slum homes, but that there was not more of it. One group of young women reported on by the Philadelphia Vice Commission showed that only a very small percentage were introduced to sex by somebody in the household.[54] Nonetheless, these households provided a considerably detailed education in sex.

Beyond this, many of those girls felt unloved, which indeed they might have been by parents utterly occupied with trying to bring home enough money each week to buy food and pay the rent, especially when it was a mother alone caring for several children. With nothing for them at home, they took to the streets as soon as they could, which was often by the age of twelve. And if they had not been educated in sex at home, they would be in the streets. "In all the cities of the state a large number of girls from twelve to eighteen years of age roam the streets at night, or frequent the parks far from their homes, rapidly learning the lessons of prostitution," the Massachusetts Vice Commission report said.[55] According to Robert A. Woods and Albert J. Kennedy, two influential early social scientists, "In many cases she takes a girl companion, with whom she roams the streets or haunts the lobbies of theaters and grill-rooms, adventuring in search after association and pleasure. It is a general opinion that young women are more and more freely entering saloons and cafes."[56] In Minneapolis it was reported that many girls were in the evening "loitering about the fruit stores, drug stores and other popular locations, haunting hotel lobbies, crowding into dance halls, the theaters and other amusement resorts; also in the saloon restaurants and the chop suey places and parading the streets and touring about in automobiles with men."[57]

With no money of their own, if they wanted even so much as a soda, much less to go into a theater or a saloon, these adolescent girls had to have a man to pay for them. Inevitably they were easy prey for men who had a couple of dollars to spend for drinks or a vaudeville show. To these girls, who spent their days in wretched, filthy sweatshops and their evenings in crowded and grimy tenements, even the seediest sort of saloon, theater, or dance hall seemed glamorous. That they would be seduced early and often was hardly surprising. On the basis of slim statistics—all that we have—it is possible to conclude that fully a quarter of the girls who later became prostitutes had their first sexual experience with a "stranger"—that is, a casual pick-up.[58] And it happened early: a quarter to a third of this population had their first sexual experience before they were fifteen, and perhaps as many as 5 percent were sexually active by their twelfth birthdays.[59] Not more than 25 percent, and probably not more than 10 percent, were virgins at eighteen. Many of these girls had been aware of prostitution from early ages, and those who had not, quickly learned about it. And at some point it was borne in on them that they were silly to be giving away what they could sell.

The idea that these girls were driven into prostitution by low wages was true then, but only indirectly, in the sense that the general effects of poverty created conditions in which prostitution had an appeal.

It should not be thought that the brothels were filled entirely with the children of immigrants. The foreign-born themselves were under-represented: various surveys put the number of foreign-born among prostitutes at 11, 27, and 89 percent.[60] High percentages of them were women of the old American stock, mainly from the farms and small towns.

Yet despite this, the *children* of immigrant families contributed "out of proportion to its percentage in the population."[61] And this was simply because these were the people who suffered most from the conditions created by the industrial city.

During the 1910s there was an enormous scare campaign, managed mainly by reformers and the yellow press which found it made good copy, about a putative well-organized white slave trade, run by foreigners, especially Jews, which shanghaied young women into brothels.[62] As the stories went, most of the women in brothels had been lured there under some pretext or other and forced to prostitute themselves. There were many tales published about girls who were drugged and woke up in sordid surroundings to find themselves deflowered. Too ashamed to return home, they accepted their fate and eventually, coarsened by the life, died of disease and were flung into paupers' graves. One writer claimed that there were 300,000 to 500,000 white slaves in captivity.[63]

It appears to be true that there existed a certain body of young men who made a practice of seducing young women, and then persuading them to become prostitutes, either collecting a portion of their wages as pimps or "cadets" as the term was then, or selling them directly to a madam for a flat price. Other men made a point of meeting immigrant ships and attaching themselves to single women on the docks. Some employment agencies notoriously sent naive girls to brothels, where they were told they would find work as domestics.[64] It is also true that some women continued in the trade reluctantly because they were afraid of their pimps, or because they were in debt to their madams, and were too ignorant to know how to get away.

But in actual numbers, the women forced to become "white slaves" were very few, if there were any. The Massachusetts Vice Commission said that it had "found very little evidence of coercion or threats of bodily-punishment on the part of these pimps or procurers.[65] The Philadelphia commission said, "No instances of actual physical slavery have been specifically brought to our attention."[66]

The simple fact is that most of these young women became prostitutes voluntarily. According to the St. Louis report, of 2,685 women queried, 2,288, or 85 percent, said that they had chosen the life.[67] And they did it not for the money; in survey after survey the girls would say that they entered the trade in order to buy nice clothes, enjoy a better lifestyle, or in a surprising number of cases, simply because they "liked being bad." But behind everything was the condition imposed upon them by the new industrial society—poverty, hard work, no comforting family to turn to for solace and support. And so they—the daughter of the immigrant, the black fleeing the harsh oppression of the deep South, the girl driven from the farm by the steady fall of commodity prices—came looking for the glamor of the big city and found instead the brothel.

This trade could only exist, of course, because there were customers. Such figures as we have suggest that the women turned about ten to twenty tricks a night. The Chicago Vice Commission obtained brothel record books which showed "Kitty" had twenty-four customers on Sunday, fourteen on Monday, twelve on Tuesday, nine on Wednesday, and seventeen on Thursday. "Florence," sixteen years old, averaged twenty-six tricks a day, and one day turned forty-five. But even this was not a record, for the Commission got testimony about a girl who turned sixty tricks in one night, which comes out to one sex act every twelve minutes for twelve straight hours.[68]

The sale of liquor was an important source of income to the madams. In many cases customers were required to buy drinks at exorbitant prices—commonly a dollar for a bottle of beer in the ordinary places, as much as twenty-five dollars a bottle for what was usually ersatz champagne in the fancy houses; in a saloon a large schooner of beer was a nickel. At these prices madams were eager to have the customers hang around to drink, and as a consequence most of these places, except for the fifty-cent houses looking for maximum turnover, provided some kind of entertainment. "In all these houses there are dance halls for the inmates and patrons where music, dancing and beer drinking are carried on," the Little Rock report said. "Many of these are elaborately furnished, and their walls are decorated with lewd pictures, which are intended to raise the passions of men."[69]

It was, however, not necessary to go to a brothel to dance. Around the brothels were clustered a variety of adjunctive establishments that were frequently more profitable than the whorehouses: saloons, dance halls, gambling joints, drug dens, restaurants, and theaters. The smaller vice districts in towns like Elmira, New York, might consist of not more than two or three brothels, a hotel or two where free-lance girls could take their customers, and a couple of dingy saloons. But in places like the Barbary Coast or New York's Tenderloin anything and everything was available. It is important to my argument that we understand that these vice districts were entertainment communities offering a whole style of life. The prostitutes were the *sine qua non*, the bait, but visitors usually spent much more of their time and money elsewhere in the district—watching a salacious show, trying to pick up women in the dance halls, drinking in a saloon. Many people came to the most famous vice districts, like the Barbary Coast or Storyville, simply to walk around and look at everything. Stride pianist Willie "The Lion" Smith hung around Newark's Coast as a boy, doing a one-legged buck and wing in saloons for small change. He remembered, "As you walked around the Coast you could always hear the tinkling of pianos from behind the swinging doors and the banging shutters of those houses they called buffet flats, or just plain cat houses."[70] (The term "buffet flat" derived from the idea that they offered a *hors d'oeuvre varié* of sexual treats.)

The saloons were often divided, a barroom in front and a second room behind for dancing and at times rough shows of one kind or another. Al Rose tells of Olivia, and the Oyster Dance, popular in Storyville:

> Completely naked, she began by placing a raw oyster on her forehead and then leaned back and "shimmied" the oyster back and forth over her body without dropping it, finally causing it to run down to her instep, from which a quick kick would flip it high in the air, whereupon she would catch it on her forehead whence it started.[71]

This, however, was a comparatively clean act. Women would copulate with animals; lesbian shows were commonplace and sometimes involved prepubescent girls; and Rose reports an instance when two twelve-year-old virgins were auctioned off.[72] Events of this kind were usually very lucrative: a woman might get a $100 to put on an act with a pony for a group of sports, and the two virgins brought $750 each, more money than many working people made in a year at the time.

Indeed, some saloons doubled as brothels. In one a balcony above the main floor contained booths overlooking the little stage below. Curtains could be drawn over the booths, so that customers could be serviced by prostitutes without having to let go of their schooners. Waiters understood that they were to stay away from the booths while the curtain was drawn.

The vice districts, then, were little communities containing a population of men as well as women: waiters, bartenders, itinerant musicians and entertainers, cadets or pimps, gamblers, drug dealers, and petty thieves who preyed on drunken men; various kinds of gangland overlords who made the districts their headquarters. The men and the women in the vice districts formed a subculture, a community within the larger culture of the industrial city. Among themselves they found their friends, enemies, lovers, even husbands and wives. Madams often served as surrogate mothers to the girls they employed. Ruth Rosen says that they often were actually called "mother" by the women, but adds that, "The madam's relationship with the inmates, however, was necessarily ambiguous and complex. As one investigator pointed out, 'Madams become the advisor and friend of the girls, while at the same time she drives them to the utmost to earn larger profits for the houses.' She was, in fact, both friend *and* exploiter to her 'girls.' "[73]

The girls depended mainly on each other for friendship and emotional support. The whole system was exploitative, and real friendship could only exist between those who performed the same function in it: madams, prostitutes, customers. Although the women were competing for men, and sometimes scrapped over the pimps, they appear to have felt a certain comradeship and loyalty to one another. Despite tensions endemic in an exploitative system, it was a true community bonded by, as much as anything, a contempt for the square outside world which looked down on them, but which came to buy from the buffet of pleasures they offered.

Yet withal, the sense of community obtaining in the vice districts was thin. For one thing, the children who lived in them were frequently neglected, with the usual results: many of them were routinely drinking, smoking, sniffing cocaine, or even gas from the street lamps, some of them as young as eight years old, the Philadelphia commission reported. "Numbers of boys in knee pants are commercializing themselves openly on our streets for the use of perverts."[74] Many worked as messenger boys, who operated out of central offices and on call made deliveries for the prostitutes, or brought them food and liquor. The "faries" (*sic*) often made approaches to the messenger boys.[75] One such boy started patronizing the brothels at fifteen, and was going once a week by eighteen. Another one went twice a week until he got a bad case of venereal disease.[76]

In Chicago a lot of boys were homeless, living in unheated sheds and under stairs. One of them, known as Red Top, hanged himself at fifteen—at least so goes the story. In his suicide note he said, "They ain't no fun living this way and I'd sooner be dead. I never had no mother or no father and no home. I don't owe nobody nothing and I don't want nobody to cry over me. So good-bye." He was given an elaborate funeral by the denizens of the district.[77]

To a startling extent, then, the inhabitants of the vice districts were young people, a huge percentage of them teenagers. It was a time, of course, when juveniles could generally leave school legally at fourteen, although of course many of them quit at younger ages to go to work; and by sixteen most people in the working class were treated as adults. Nonetheless, the exploitation of the young in the vice districts undercuts any romantic notions of glamor.

It was basically a hard world, where drug addiction, alcoholism, disease, violence, and suicide were commonplace. Yet they were human beings and they could have fun. The Hartford Vice Commission report described an evening in a wine room used by prostitutes. "Eight men and five women were smoking, singing and indulging in suggestive dances. Couples frequently left the room and returned in about half an hour, and the remarks made upon such occasions plainly indicated the nature of the occupation while they were away. None of the girls in this room were over twenty. The 'fun' continued from 4 until 11 p.m." [78]

It should be made clear that prostitution, despite unofficial efforts to restrict it, was hardly confined to the vice districts. The Philadelphia report said that "in saloons, cafes, restaurants, hotels, clubs, and dance halls prostitution is being constantly engendered and fostered . . . Many public dance halls, moving picture shows, and other amusements are the breeding-places of vice—the rendezvous of men who entrap girls and girls who solicit men." [79] One report estimated that there were five to ten times as many women working outside the district as within it, [80] and other estimates substantiate these figures.

However, it was the red light district as an institution which had a major impact on the culture of the United States. The free-lance prostitute offered sex for money; but the vice district presented a way of behaving, an attitude toward life that was almost diametrically opposed to the Victorianism that was still, in the period of about 1880 to 1910 when the vice district was in its heyday, the established—indeed almost official—code of the United States. Victorianism favored the abolition of drink and drugs, sharp curbs on sexuality, the end to gambling and pornography. But the vice districts were in the business of selling exactly those commodities—drugs, liquor, dirty pictures, gambling, and sex in every imaginable form.

The Victorians believed deeply in the sanctity of children and the family. Toilers in the vice districts, who usually lived outside of families, did little to protect the children in their midst from exploitation by adults, or indeed other children. Finally, the Victorians believed that a duty was owed to the community and that sacrifices must always be made for the general good. The denizens of the vice districts had no concept of duty, and the only sacrifice any of them made for the common good was to avoid the more extreme activities, like murder, which might discourage trade.

The vice district, then, was antithetical to the Victorian morality still being upheld by the middle class well into the period of World War I. It was in the business of purveying self-gratification. However, not only was it selling self-gratification, it was also living by the ideal of self. Despite the loyalty that sometimes obtained between the prostitutes, or the women and their cadets, there was among these people little thought for the morrow, little concern for the larger community, little interest in the welfare of those around them. They had come into the district—the saloon-keepers, waiters, prostitutes, brothel-keepers, police, landlords, and all the rest—solely to earn money and indulge whatever sensual appetites their fancies suggested. Basic to the code of the vice district was the cult of the self.

Into these districts regularly came men, and some women, from all sections of the American society. We are once again lacking in statistics, but it is probable that the majority of the men who visited vice districts frequently were drawn, like the prostitutes, from the working class. Kinsey figures show that working-class males tend to have much more intercourse with prostitutes than do middle-class males, and some of the people in his sample were young males when the vice districts still existed.[81] There were, of course, good reasons for the preponderance of working-class males in the vice district. For one thing, there were three times as many of them as white collar males; for another, as most of the vice districts were established in working-class neighborhoods, they were handy and well-known; for a third, the male-female imbalance which existed after 1900 was largely a working-class phenomenon.

But middle-class males had one very good reason for visiting brothels: lack of sex partners in their own social class. Working-class males usually had among them at least a few women—or girls—who were available for sex. These were the girls the vice commission reports were full of, who were haunting the hotel lobbies and dance halls in search of fun, or even something to do. But the younger middle-class male simply did not have in his own social group *any* young women who were sexually available. In those days, and for many decades after, middle-class women simply did not have sex before marriage—at least not until they were well in their twenties, and most not even then. The girls these boys went to school with, took to country club dances, met at the houses of their parents' friends, fell in love with, could not be touched—not even kissed in many cases—before they were engaged.[82] If these young men were to have any sex lives at all, they had to find it outside their own social groups; the vice districts stood ready.

What percentage of middle-class males actually visited the vice districts is difficult to arrive at. The Kinsey figures are at times hard to pin down, because he did not break the statistics for males down into cohorts, but only into two groups: an older generation born on either side of 1900, and one born some twenty years later. Extrapolating from his figures—which requires making an educated guess as to who belongs in which social class—it appears that something like 40 percent of the middle class visited prostitutes in the heyday of the vice districts. Many of them paid for sex several times a month, and perhaps the majority of single males did it once a month by their mid-twenties.[83]

Other figures suggest that about a quarter of male college students visited a prostitute at least once as undergraduates.

Not all of these young men were going to the vice districts, of course; many of them were frequenting the free-lancers. But it is clear that substantial numbers of middle-class males did visit the vice districts, including numbers of married men. The Pittsburgh report insisted that "married men make up a very large proportion of the frequenters of the red-light district."[84] The Elmira report listed among the clients to one brothel on a certain evening four "elderly farmers," three railroad men, a group of streetcar conductors and motormen, and a sailor. "Most all were drunk—a boisterous, tough crowd." But among them were "a well-known citizen," and "two swell fellows."[85] The New York reformer Charles Henry Parkhurst, who visited a number of brothels in the course of his campaign against vice, said that "our best and most promising young men" were found in the districts.[86] The Massachusetts report said, "It can be said that the men who are frequent customers in these resorts come from many walks and occupations, from the loafer on the street corner with no visible means of support to prosperous looking clerks, mechanics, business and professional men."[87] And one Washington, D.C., observer wrote:

> And whenever baseball or football games brought hundreds of collegians to the national capital, scenes quite similar to those of inauguration nights were to be witnessed. Washington has a fairly large student community for its size and the effect on their lives with a district of prostitutes always available to them can very well be imagined. It became a custom for these young men to saunter down after the theater to the houses along Ohio Avenue, D Street, and the like, in the northwest. Groups dashed boisterously thither in automobiles for merry-making into the wee hours of the morning . . ."[88]

These young college men with their automobiles were the American upper crust, the top 10 percent of the social structure. They had both the money and the leisure time to spend on this kind of adventuring, and many of them did it frequently.

Nor was it just the males: a certain number of middle-class women occasionally visited the vice districts. Herbert Asbury reports that in the last years of the Barbary Coast, which was shut down early in the twentieth century, Anna Pavlova and Sarah Bernhardt visited the district,[89] and other sources speak of parties of "slummers" who visited the Barbary Coast dives in the latter years.[90] One of the points of the vice districts was that they were visible in a way that the solitary prostitute was not—or not to the degree.[91]

An inexperienced young man from the middle class might not know where to find a free-lance prostitute: they usually did their business from saloons, hotel lobbies, and street corners that such a man might not usually frequent. But the vice districts were known to everybody. "The character and business of these parlor houses are generally well understood by the neighbors—indeed, by the whole community in the smaller cities and towns," the Massachusetts report explained. "Information concerning their location is freely given by cabmen, streetcar conductors, and others."[92] Word of the districts circulated among groups of boys when they were quite young, and indeed, almost everybody in the United States with any pretense to sophistication knew about the famous places like Barbary Coast and Storyville, which was, after all, legal for twenty years.

And what these middle-class men found in the vice districts was not just sex: it was a whole style of living, coupled with an attitude that justified the lifestyle. Whatever the sellers of sensation actually felt about the trade they were in, they perforce presented it as right and legitimate. The vice districts, in sum, *institutionalized* anti-Victorian attitudes and behaviors. In so doing they provided a school for the American middle class in a way of thinking, feeling, and behaving that had not been part of the culture they had been raised in.

CHAPTER 6

The Victorians Fight Back

The new industrial city was essentially anti-Victorian. Its character was determined not by the Victorian middle class that still resided in it, but by the rough culture of the working people, with their confused, boiling ethnic mix. Working people were the majority; they were physically present in the streets in a way the middle class, sheltered in homes, offices, schools, and shops, was not; they rapidly, through the first years of the twentieth century, pushed their way into political offices; and they were, in those same years, creating, or having created for them, the most characteristic institution of 20th-century America, the huge entertainment industry which has come to so largely occupy American lives. In the new city, the tone was set by the working class. And that tone was not a Victorian one.

But the middle class, in the early years of the 20th century, was still firmly and deeply committed to the Victorian idea of family and church, decency and order, gentility and cleanliness; it was still trying to do with ease what it did not want to do. In sum, it was urging control of the self for the general good. It seemed self-evident to these people that this ideal was the only one that made sense, and they were, many of them, horrified by what they were seeing around them in exploding cities—the filth, the squalor, the open sexuality, the carousing. Inevitably, they began to fight back. And through the last decades of the 19th century and the first decades of the 20th they mounted a series of attacks on alcohol, sex, dirt, crime, corruption, even sports and the tendency of the unwashed not to wash.

This Victorian backlash manifested itself in a series of movements— indeed crusades—expressed through reformist organizations. These movements were not necessarily related to one another, and in some cases were contradictory; but surprisingly frequently they were coordinated, and in many cases actively supported one another, or indeed had overlapping personnel. For the most part such organizations were private, and depended on contributions for their support, much of it from the rich. But they were political, in that most of them aimed at getting legislation passed which would curb this or that unacceptable sort of behavior.

The best known of these reforming movements, and the most successful one, was the long fight for control over the drinking of alcohol. The anti-alcohol drive, which had a strong female component, was aimed primarily at the drunken father who drank up his wages in the tavern, neglected his children, and beat his wife. The movement began in about 1830 and went in fits and starts, pushed along by waves of enthusiasm which rose and fell. By 1855 thirteen states had "Maine laws" which limited the sale of liquor.[1]

This anti-alcohol movement was to a considerable extent anti-immigrant. The Germans with their beer gardens and liberal attitudes toward drink were seen as bringing to the country a dangerous habit. But the anti-alcohol drive gained substantial force in the days of the new immigration of the 1870s and 1880s. The saloon, a product of the new industrial society, was looked upon in horror by the reformers, with some justification. However, large numbers of working men took a different view: many of them were living as lodgers in cramped quarters, and the saloons provided them a place in which to socialize in their spare time. Roy Rosenzweig, in his study of working-class leisure pursuits, says that the majority of males in Worcester, Massachussetts, went to the saloons every night.[2] Probably the saloons were neither as bad as the reformers believed, nor as innocuous as the brewing interests claimed.

The big brewers eventually owned 70 percent of the saloons[3]; with tens of thousands of them everywhere in the big cities, competition was fierce, and proprietors sold alcohol to anybody who had money, regardless of age and condition of sobriety. "The case against the saloon, the drunkard, and the liquor traffic rested on the social statistics of crime, delinquency, poverty, prostitution, disease and corruption."[4]

The founding of the Anti-Saloon League, which had its first convention in 1895, was critical in the drive against alcohol. It was, according to one student of the subject, Norman H. Clark, "the most effective pressure group in American political history."[5] The focus on the saloon as the enemy was the key, for it seemed that the saloon was a noxious, indeed destructive element in American life, and few people of sense could find much justification for it. Making matters more difficult for the wets was the fact that the brewers ran ineffective campaigns.[6] By 1900 thirty-seven states had local option laws[7] which allowed communities to do what they liked about alcohol, and quite rapidly the country began to dry up. Oklahoma went dry in 1907,[8] and by 1916 a majority of states had some kind of prohibition laws on the books.[9] The game was over, and in 1920 the Prohibition Amendment went into effect. Once again the Victorians appeared to have won.

Prohibition, says Joseph R. Gusfield in his study of the question, was "a middle class, Protestant, and nativist activity."[10] It is doubtful that at any time a majority of Americans really wanted to eliminate alcohol from their lives altogether. What most people wanted to curb were the excesses—the noxious saloon, the alcoholic working man who drank up his pay on Saturday night and went home to beat his wife, the selling of liquor to minors, and the like. The prohibition laws went through primarily because the rural old stock, the strongest anti-alcohol population in the country, was able to dominate state politics. The immigrants, by the 1910s, controlled the cities to a considerable extent, but the rural districts controlled everything else, and for the moment they had the power.[11]

A second thrust in the Victorian backlash of the late 19th century and early 20th was what was called "the purity crusade." It began mainly in opposition to the movement for regulated prostitution which in 1870 resulted in the creation of the St. Louis segregated vice district. The argument for regulation was that males had a sexual drive they found difficult to control; that it was better that they should have recourse to prostitutes than to debauch decent girls; and that an organized system of prostitution, which could be kept orderly and disease free, was preferable to the uncontrolled system which then obtained.

The purity crusaders responded that the males were perfectly well able to control their lust, and that it was merely self-indulgence that led them so regularly to prostitutes. They insisted that the double standard be abandoned, and that males should be held to the same rules of sexual abstinence enforced on women—that is to say, sex should be confined to the marriage bed.[12] The purity crusaders, furthermore, felt a concern for the prostitutes themselves, whom they saw as victims of male lust. A great deal of effort was expended on rescuing the women from the brothels, most of it wasted on girls who had gone into the life voluntarily and preferred it to the sweatshops.[13]

But as the purity crusade gained momentum, it swept into itself other causes. David J. Pivar, in his study of the crusade, says, "Initially only a humane interest in rescuing prostitutes and suppressing reglementation [i.e. regulated prostitution], purity reform became a symbol of a totally purified society . . . reformers moved from a defense of traditional morality to an aggressive attempt to control the social environment."[14] By the 1890s the purity crusade, Pivar says, "had unquestionably become a mass movement."[15] In particular, the purity crusaders saw the new industrial city as an abscess. "By the 1890s, reformers, alarmed over crime increases, distressed over the widening social distance between the wealthy and the poor, dismayed over moral disintegration, and discouraged by the social irresponsibility of America's

wealthy families, came to see American cities as diseased."[16] They began to go after everything—boxing, gambling, football, ballet featuring
tights, nudity in general.[17] After 1895 it became an attack on the new
mass entertainment system, another artifact of the industrial city, with
assaults on movies, the recording industry, the penny arcades.[18] It even
included a fight for cleanliness that urged the building of public baths
so the unwashed could bathe.[19] The purity crusade seemed to oppose
almost any kind of pleasure, and especially the pleasures of working-
class males.

One of its most significant effects was the abolition of the segregated
vice district. The purity forces were well aware that the police and other
municipal authorities were conniving at the existence of the vice districts, and they knew perfectly well that huge sums in bribes were being
paid by brothel owners and adjunctive establishments annually for per-
mission to operate. It was clear to them that the only way to close down
the districts, and with them prostitution, was through the election of
reform governments. They had in their favor certain critical facts which
were apparent to everybody. The first was that the segregated vice districts did not segregate: one vice report after the next concluded that
the largest number of prostitutes were operating outside of the demar-
cated districts. The Little Rock report, for example, estimated that there
were four to ten times as many women working outside the district as
inside it.[20] (Such estimates, of course, supported the reformers' case.)
For another thing, it was also clear that segregated prostitution did
little or nothing to reduce the incidence of venereal disease. The primary reason was that there were few inspections, and fewer sanctions
against those found infected. The consequence was that various vice
commissions found plenty of disease in the segregated districts—in Baltimore only 3.4 percent were uninfected,[21] and in New Bedford, Massachusetts, 90 percent were ill.[22]

Faced with these facts, it was very difficult for anybody to make a
legitimate case in favor of the districts. Very quickly in the years around
1910, when the reform movements were peaking, city after city put
through strong laws abolishing the vice districts. For example, in 1908
Iowa passed a law which called for the ouster of officials who were
winking at the red light districts and "the day after the laws in regard
to the social evil went into effect cities which had had open houses of
prostitution for fifty years found them closed."[23] It was, in fact, a rout:
in the second decade of the century "over two hundred cities have closed
their districts."[24] The last of the major vice districts to close was Storyville, shut down by the Navy Department over the city's protests in
1917.

A third strain in the Victorian backlash of the decades on either side
of 1900 was what has been called "progressivism." This was basically an
effort to reform American society through the election of educated "so-

cial planners" and the passage of legislation to cure the most serious ills which the middle class in particular saw around it.

But there was more to it than politics, for it was as much a philosophy or at least an attitude as it was a set of political ideas. Richard Hofstadter said, "The country seems to have been affected by a sort of spiritual hunger, a yearning to apply to social problems the principles of Christian morality which had always characterized its creed but too rarely its behavior. It felt a greater need for self-criticism and self-analysis."[25] It was triggered in considerable measure by the wave of strikes and the general industrial disorder of the 1880s and the 1890s, coupled with a growing awareness of the corruption in government and business. Muckrakers like Ida Tarbell and Lincoln Steffans were writing tough descriptions of wrongdoing by corporations and bribe-taking by politicians. Upton Sinclair's *The Jungle,* detailing the appalling conditions in the Chicago stockyards, was widely read. "Outpourings of anger at corporate wrongdoing and hatred for industry's callous pursuit of profit frequently punctuated the course of reform in the early 20th century. Indeed, anti-business emotion was a prime mover of progressivism,"[26] say Arthur S. Link and Richard L. McCormick. Theodore Roosevelt said, "The dull, purblind folly of the very rich men, their greed and arrogance and the corruption in business and politics, have tended to produce a very unhealthy condition of excitement and irritation in the popular mind, which shows itself in the great increase in socialistic propaganda."[27]

Progressivism, then, was a broad movement that encompassed many aims and reforms. "Progressivism was the only reform movement ever experienced by the whole American nation,"[28] say Link and McCormick. They add, "In general, progressives sought to improve the conditions of life and labor and to create as much social stability as possible. But each group of progressives had its own definition of improvement and stability."[29]

The key was, as Robert Wiebe has said, "a government broadly and continuously involved in society's operation."[30] To that end the progressives pushed through city, state, and the national governments an array of reforms. But the progressives did not confine themselves to politics. By 1910 they had created 400 settlement houses for the integration of immigrants into American society;[31] pushed through the construction of all sorts of recreational facilities, like playgrounds, based on the idea that much of the immorality of the ghettos arose because the young there had little by way of decent amusement;[32] involved themselves in the anti-vice crusade; and pushed through pure food and drug laws and housing standards to eliminate the squalor of the slums.

Progressivism, clearly, subsumed a lot of the goals of the anti-saloon movement, and the purity crusade. "Year by year the innovating spirit whirled faster and faster, overrunning old objectives, generating new schemes for reform, broadening into a force that stirred every nook and cranny of American life. By 1910 the excited talk of reformers about 'social democracy,' breaking the 'money trust,' shackling the judiciary, and emancipating the 'new woman' suggested that a reconstruction of some of the country's major institutions from the banking system to the home, lay in the offing."[33]

The progressives thought of themselves as modern, an advance guard applying the tenets of the new social science to the problems of democracy. And it is certainly true that their scheme of using the government to intervene in the workings of society for the general good laid the philosophic basis for the New Deal, the Fair Deal, the Great Society. But paradoxically, as many recent historians have insisted, the progressives were essentially conservative—Victorians at heart. In their desire to improve conditions in the slums they were espousing a communitarian ideal. In their support for child welfare, child and female labor laws, they were attempting to promote the interests of the family. In their efforts to improve sanitation and regulate housing, they were calling for the decency which was so critical to Victorianism. In the building of parks and playgrounds, they were trying to subdue chaos in the streets in pursuit of a more orderly society.

It should be clear that the Victorian backlash combined a genuine concern for the ills of the American "community" with a desire to restrain behavior in general in the interests of order and decency. Many in the late Victorian middle class would not have made this distinction, however. They would have seen both child labor laws and anti-alcohol legislation as contributing to the general welfare of the American people.

When we put the strains of the Victorian backlash together, we can see that its purpose was to replace the old idea of self-constraint with legal restrictions. In an early day the Victorian male, it was hoped, would voluntarily limit his drinking, curb his lust; the Victorian female would put herself at the service of her family and her community through voluntary do-good societies. But the new big city working class, especially the immigrants, did not seem to be willing to impose such constraints on itself, and therefore there would have to be laws imposing a measure of selflessness on the citizenry.

It should not be thought that the middle alone was concerned with social issues. Roy Rosenzweig says that "the temperance crusade was, in part, an effort by the city's middle and upper classes to reform, reshape, and restrict working-class recreational practises," but he goes on to point out that there was always a minority in the working classes who wanted to curb excessive drinking.[34]

Still, the temperance movement was primarily a middle-class cause, and it is not difficult to see why the reformers of the Victorian backlash came to be seen as crabbed, pinch-faced Puritans unable to take pleasure in anything, and determined to see that nobody else did either. But we must remember that indignities and vices they saw were real. Twelve-year-olds *were* being seduced by grown-up men; laborers *were* drinking up their pay in saloons and going home to beat up their wives; parents *were* forcing their children into the sweatshops at the earliest possible moment. Probably the majority of the people who provided the soldiery for these crusades were not scaley Puritans, but ordinary people who saw misery and want all around them and hoped to improve matters.

And it appeared, by 1912, that the reformers were succeeding. By that year they had put thousands of people into public office; the votes for the left-wing parties soared;[35] state after state was shutting down the breweries; city after city was closing the brothels; parks and playgrounds were springing up everywhere; and a flood of reformist legislation was rolling through state houses.

But the triumph of the reform movement was illusory. Americans began turning against Prohibition almost from the moment it was passed; a new sexual morality, the first springs of which were visible by the time of World War I, would make the fight against the vice districts irrelevant; a new conservatism following the end of the war would vitiate the progressive innovations.

What really happened was that the onward roll of the new industrial society had gained enough momentum that it had become unstoppable. The Victorian backlash was in fact failing just when it seemed to be having its greatest triumphs. Some people saw this. As progressivism was peaking, George Santayana said, "The Civilization characteristic of Christendom has not yet disappeared, yet another Civilization has begun to take its place."[36]

What happened? Why, when virtually the whole middle class, as well as a considerable portion of the farmers in the rural areas and the working people from the old stock, wanted to keep in place the Victorian idea of subordinating the self to the family and the community, however modified, did it disappear? The answer, as I hope I will be able to show, is that the middle class was simply seduced.

CHAPTER 7

The New Business
of Show Business

The industrial society which was coming into being in the United States in the 1880s and 1890s produced a number of new institutions, like ward politics, the vice district, the ethnic slum, all of which have shown varying degrees of staying power. But it was another institution arising out of the industrial city that would go on to dominate American society and eventually the international culture which came into being after World War II. That was the organized entertainment industry, today a huge international business and, not surprisingly, one of the few industries in which the United States continues to play the major role. Americans today spend more of their time being "entertained" than they do anything else, including working and sleeping. It is a major component of their lives, not merely in terms of the time they spend at it, but in the extent to which they are emotionally involved with their favorite television shows, musical heroes, sports stars, and those rather generalized "celebrities" who are famous primarily for being well-known. And what is true of America is rapidly becoming true of other nations in the industrialized world. The 20th century is truly the Age of Entertainment.

In colonial America, and well into the Federal period, professional entertainment was sporadic and in short supply. With 90 percent of the population scattered across the countryside in isolated farms and for the most part without any transportation but shoe leather, it was very difficult to gather an audience sufficiently large to make a performance pay. Furthermore, living to an extent by barter,[1] many people would have had to pay for their tickets in bushels of apples and sacks of cornmeal, which traveling entertainers would have found awkward to deal with.

There was a good deal more entertainment available in cities and to an extent in small towns. New Orleans had its so-called "French" opera house from 1796, and in the 1805–6 season produced sixteen operas.[2] Shakespeare, frequently mutilated, was very popular in early 19th-century America,[3] and many cities offered concert halls, waxworks, menageries, and "museums"—mainly depositories of curiosities.

But even in the cities the amount of formal entertainment available was limited. For one thing, a residual Calvinism still held the theater to be sinful. According to A. H. Saxon, in a recent biography of the showman P. T. Barnum, a "hatred of [traveling] entertainers," dating back to the English Restoration, continued well into the 19th century, and even beyond.[4] This view was reinforced by the character of much entertainment of the day. Saxon says, "Shady practices and loose behavior on the part of employees—not to mention the coarse jests and indelicate references that figured in many clowns' repertories—frequently outraged the moral element in communities; and travelling establishments, especially, like carnivals in a later day, were often the scenes of bloodshed and violence, frequently brought on by local roughnecks who viewed all such intruding 'showfolks' as beyond the pale and thought nothing of killing an innocent performer for their added amusement."[5] Paul E. Johnson, in his study of Rochester, New York, for the period, says that at one performance of *Othello* in about 1830 the manager "had to stop the show and plead with the audience to stop shouting and throwing things at the actors." One Rochester newspaper editor complained that when a show was on the uproar was so great he could not sleep at night.[6]

By the 1830s and after, the amount of entertainment that was reaching the hinterlands was growing, especially as towns began to link themselves together with roads. A few opera troupes were touring; circuses, mainly consisting of menageries and displays of horsemanship,[7] were becoming increasingly common; and in 1831 William Chapman built the first of the riverboat theaters—the famous "show boats"—which roved the major rivers, pulling in at villages along the way.[8]

Perhaps the most important of these institutions were the so-called museums, which by mid-century existed in most American cities. The earliest of these was the one organized in Philadelphia in the 1780s by Charles Willson Peale, a naturalist and one of the best-known American painters of the time. According to Saxon, Peale's museum contained a menagerie, collections of minerals, insects, seashells, animal skeletons, a living cow with five legs and two tails, portraits, and wax figures. It offered, from time to time, lectures, musicales, and scientific demonstrations.[9] There was a similar museum in New York in the 18th century, and others elsewhere. The point was that a museum was not there to entertain but to edify, and it therefore did not trouble the consciences of residual Calvinists, or the new Victorians.

The key figure in the museum business, however, was Phineas T.

Barnum. Barnum began his career in show business by touring a black named Joice Heth, who was purported to be 161 years old and George Washington's childhood nurse.[10] Eventually Barnum bought the American Museum in New York. Over the years he expanded it, turning it into a colossal whatnot cabinet which included animals both living and dead, an aquarium containing some small whales, and floor after floor of exhibitions of every conceivable kind.

Barnum eventually lost the museum in a fire, but Saxon has calculated that it was, in its time, relatively more popular than Disneyland in its.[11] Barnum, of course, is best remembered today for the circus which still bears his name, but he did not really get into the circus business in a big way until 1870, when it already had a long history.[12]

In any case the number of traveling shows available to the farm and village families who constituted the bulk of the American population was small until, really, the 1880s, if not later. It is difficult to put figures to anything like this, for there must have been hundreds, if not thousands, of traveling performers moving around the United States in the middle decades of the 19th century who went unrecorded. But in *Huckleberry Finn,* Mark Twain's portrait of small-town America in the ante-bellum period, it is clear from the famous "Duke and the Dauphin" episodes that Huck, then about twelve years old, had had little experience with circuses or shows in general. Few farm families had either the time or the money to take in a traveling show very often; and during a hard northern winter, not many would have been much inclined to hike three or four miles through the snow to enjoy one. For most Americans during the colonial period and for decades after, professional entertainment was a relatively rare treat. Like butter and candles, American entertainment was usually home-made—a dance, a footrace, a family sing around the square piano or cottage organ in the homes of the better off.

Among the few traveling shows which did meander around the United States were some entertainers, working by themselves or with small troupes, who did blackface acts to give what purported to be imitations of Negro song and dance. The most famous of these entertainers was Thomas D. Rice, who copied a dance he saw done by a black man stiff with age and crippled with rheumatism. "Weel about and turn about and do jus so; ebery time I weel about, I jump Jim Crow," the old man sang, and Rice's version of the dance became widely popular in the 1830s.[13]

In time these blackface acts were expanded, with larger troupes playing more elaborate shows. This form of entertainment came to be called the minstrel show. It was made up of set routines combining songs, dances, jokes, and even fairly lengthy skits, all supposed to represent the jolly life of blacks on the old plantation. Minstrelsy "swept the nation" in the 1840s,[14] and remained popular for decades thereafter. In

the post-Civil War period, some blacks, now able to travel and to or-
ganize their own businesses, put together minstrel shows of their own,
ironically blacking up to maintain the character of minstrelsy.[15] These
black minstrels began the tradition of the black entertainer in the United
States, which has played so profound a role in American entertainment. Nonetheless, whites continued to dominate minstrelsy.

By the 1860s and 1870s a new, although parallel, form of entertainment was emerging which would eventually kill the minstrel show and
go on to become the basis for the modern entertainment industry. This
was variety. Like minstrelsy, variety consisted of a loose collection of
songs, dances, comedy, skits, and other hard to categorize acts; but it
was not tied together by a common theme, as minstrelsy had been.
Variety owed something to minstrelsy, in part because the minstrel shows
had worked out the system of moving a large troupe from town to
town, and in part because variety shows frequently included black acts
which had been developed by the minstrels.

But the true roots of vaudeville lay in Europe. Beginning in about
the 1840s there had developed in European cities the institution of
what was called in France the *café-concert*. Often set in the open air,
these *cafés-concerts* consisted of tables grouped around a stage where
fairly rough entertainment was put on. Waiters went through the audience serving liquor, mainly beer.[16] One important feature of the *cafés-
concerts* was the mingling of social classes, as we can see in Édouard
Manet's "The Waitress," which shows a top-hatted gentleman seated
close to a blue-smocked working man in a *café-concert*.[17]

The immigrants pouring into the United States from the late 1840s
on brought the idea to America. Kathy Peiss, in her study of working
women's amusements at the turn of the century, says, "In the 1850s
some saloon owners converted their back rooms and cellars into small
concert halls and hired speciality acts to amuse their patrons and encourage drinking. By the 1860s over two hundred concert-saloons had
spread along Broadway, the Bowery, and the waterfronts, catering to a
heterogeneous male clientele of laborers, soldiers, sailors, and 'slum-
ming' society gentlemen. The conventions of polite society were put
aside in these male sanctuaries, where crude jokes, bawdy comedy
sketches, and scantily clad singers entertained the drinkers."[18]

The idea of providing entertainment to attract and hold drinkers is
an obvious one; but the timing of the rise of the concert saloon, the
mingling of social classes, and the name itself strongly suggest the European origin.

The shows might open with a chorus line of women in revealing
costumes, after which would come comics and song and dance acts.
Performances were often closed by an "afterpiece," usually an erotic
and partly improvised skit; among the most famous of these were "The
Book Agent" and "The Bathing Girl," both short on plot but long on

innuendo.[19] In most of these places "waiter girls," in short uniforms, served drinks and sometimes themselves. By the latter decades of the 19th century there were saloons of this type in all big cities and in many small towns, like Sherwood's Mascot in Galveston, Texas, and Chicago Joe's Coliseum in Helena, Montana. The most famous of these places were the big city saloons, like Harry Hill's on Houston Street, in New York City, which offered boxing matches, walking contests, song and dance, and the usual blue skits.[20]

Variety was created by moving this saloon entertainment into theaters. "At first the changes were superficial and, although more elegant than the saloon, the fare was still quite low and vulgar," employing the usual sexual jokes and dancing girls showing a good deal of flesh. Ethnic humor, built on stereotypes of the German and Irish immigrants coming into the country, was a staple of the new variety show.[21]

These early popular theaters, it should be noted, had a strong sexual component, off-stage as well as on. Many of them had third tiers above the dress and family circles reserved for prostitutes. "By the 1830s and 1840s, the relinquishing of the third tier to prostitutes had become an established national tradition," says one writer.[22] There was often a bar on the third tier, and a separate set of stairs leading up to it from a side alley. One authority has estimated that as late as 1875, 70 percent of Americans associated the theater with sin; but up until the 1880s theaters could not have survived without the patronage of the prostitutes, not for their admission fees, for they were generally admitted free, but for the men they drew.[23]

The theater, then, was not a place where women and children of the new middle class the industrial society was spawning, with their Victorian ethic, could possibly go; and this in turn meant that middle-class males could not visit them either, at least on those occasions when they wanted to take their families or sweethearts out for an evening of fun. The theaters, thus, were off-limits to the most affluent 25 percent of the American population; and it occurred to one showman, Tony Pastor, that if he could produce clean variety, he could attract a whole segment of the population to his shows which had hitherto not come. In 1881 Pastor opened his Fourteenth Street Theater, offering "a straight, clean, variety show—the first—as such—ever given in this country. It was a daring venture. Only gals on the trampish side attended variety in the eighties. Pastor's move was mainly (and frankly) for profit, a definite and canny bid to double the audience by attracting respectable women—wives, sisters, sweethearts—"[24] says Douglas Gilbert in his history of the institution.

Pastor's innovation was successful, and very quickly other entrepreneurs leapt in to imitate him. Through the 1880s and 1890s variety grew at an astonishing pace. Across the nation, barns, warehouses, abandoned churches were converted to variety theaters. Very quickly, in order to shed the old unseemly image, the name was changed to vaudeville, a word of French extraction whose origins are in dispute.

The heyday of vaudeville ran from about the mid-1890s to approximately 1920, although it had begun earlier, and was still staggering along into the 1930s. Through its golden age the houses grew more elaborate, the acts more polished and professional, the audiences ever larger, the salaries higher, and the renown greater. Douglas Gilbert says, "The essence of American vaudeville was comedy despite [Edward] Albee's contention that it was women's backsides."[25] The comedy, especially in the early days, was often very heavy-handed, depending on fright wigs, slapshoes, and the punching about of one comic by the other. But the routines were fast-paced, played with energy and zest, and followed each other in rapid fire, so that audiences hardly finished applauding one act when the next was pouncing at them.

It began to be recognized that certain acts extracted more applause than others, and promoters and theater owners started advertising, or "billing," such acts more prominently. Very soon a hierarchy of vaudevillians was created, with the stars at the top barely deigning to speak to the humble unknowns at the bottom. The top acts began demanding special treatment in the form of large dressing rooms and other amenities, and of course salaries ranging upwards of $5000 a week.[26] Even fairly low level performers could make good incomes in the vaudeville heyday.

With so much money rolling in, the acts were dressed up with stylish backdrops and used elaborate props. In order to keep the Victorian middle class coming, a certain refinement was allowed to creep in. In the late 1890s violinist Edouard Remenyi, "the Heifetz of his day," performed in vaudeville playing "Hearts and Flowers," Mendelssohn's "Spring Song," and similar works.[27] B. F. Keith, co-owner of the Keith-Albee chain, the largest vaudeville group in the country, began posting signs in his dressing rooms warning performers against the slightest impropriety. But not all performances were elegant. The cat piano routine involved a man who miaowed the "Miserere" while pulling the tails of cats imprisoned in wire cages. One dancer came on stage naked to the waist, with eyes painted on his nipples, a nose below, and mouth around his navel; by working the muscles of his torso he made faces on his stomach as he danced. Another actor did soliloquies from Shakespeare, playing Hamlet with a beard and tights, while a comic played Yorick with a German accent, and dug beer bottles from a grave.[28]

Through the 1900s and 1910s vaudeville continued to expand. "By the teens there were more than one thousand theaters playing standard vaudeville acts and in excess of 4,000 small-time theaters." One authority says that there were between ten and twenty thousand acts competing for the work,[29] but, in fact, it could not have been less than twenty thousand, and may well have been double or triple that number. The money roared in. In 1893 Keith and his partner, Edward Albee, opened the first real vaudeville "palace," the Colonial in Boston. It cost some $670,000 for the decor alone. In 1922 the chain added the Cleveland Palace at a cost of five million dollars. There were paintings by Corot and Bougeureau in the marble lobby, and the lobby carpet was the largest single piece of weaving in the world; or so the publicity went.[30]

But by this time B. F. Keith was dead, and so was vaudeville, although nobody quite realized it yet. The peak had come in the years just before World War I; the downhill slide went faster and faster through the 1920s, and when the famous Palace in New York, the nation's premiere vaudeville showcase, was converted to a movie house in 1932, the dying business stopped breathing.[31]

Vaudeville constituted the first organized system of mass entertainment in the United States—or indeed the world. Before it, entertainers had been individual entrepreneurs who worked as singles, or as small opera or minstrel troupes, playing riverboats, small theaters, and the free-and-easy saloons. But vaudeville came to be dominated by a handful of showmen who owned huge chains of theaters, and ran them with the attention to detail characteristic of big business. By the 1920s Keith-Albee had four hundred houses, and the great chains of Marcus Loew, F. F. Proctor, and two or three others had hundreds more.[32] The chains operated like network television, "broadcasting" the same acts, even entire bills through the system, the only difference being that network television is instantaneous, while it might take an act months to work its way around a major circuit. Bills were not thrown together haphazardly, but were carefully worked out to provide pace, rhythm, and variety. Acts were expected to be thoroughly rehearsed and carefully polished. There was nothing slapdash or improvised about vaudeville in its mature stages. The theater owners were by-and-large canny and usually fairly cold-blooded showmen, who were, like other businessmen, mainly interested in money. The system was operated by bureaucracies of experts in one phase or another of the business. It ran with machine-like smoothness, oiled by large sums of money, and was capable of crushing people who stood in its way.

Vaudeville was the foundation on which the 20th-century entertainment business was built. It provided a model for a national system. It turned over to later forms, like the movies, an infrastructure of theaters into which they could easily slide. It left a legacy of acts which would be used by newer forms, perhaps modified, but still essentially employing the old routines: Eddie Cantor, the Marx Brothers, George Jessel, Al Jolson, Burns and Allen, Abbott and Costello, Jack Benny, Fred Allen, W. C. Fields, and many others trained in vaudeville went on to become headliners in movies, radio, and television. The Marx brothers, Abbott and Costello, and W. C. Fields movies were built around the characters and routines they had developed in vaudeville; the Jack

Benny, Fred Allen, and similar radio programs were basically vaudeville bills, consisting of comic patter, songs and skits, and in acknowl-
edgment of their roots were called variety programs; even as late as the
1960s one of the most popular television programs ever was an out-
and-out vaudeville show led by a former Broadway newspaper colum-
nist named Ed Sullivan.

The modern entertainment industry came out of vaudeville, and
vaudeville could not have grown to the giant it became outside of the
industrial city. In its maturity it was a very expensive operation and for
its audience needed a large population within easy distance of its theaters. It needed crowds of people, it needed rapid transit systems, it
needed people with a little money in their pockets and leisure time to
spend the money on. A national vaudeville system could not have been
built on small communities with the populations of farm families. It
needed cities.

Among other things, the fact that the new industrial society was actually tied together physically by a vast metal web of railroad tracks and
telegraph and eventually telephone wires allowed vaudeville to develop
into a national system to a degree that would not have been possible
had it had to depend upon canal barges and stage coaches for transportation and the mails for communication. It was this network that
allowed industry in general to become national, and in that sense
vaudeville was "industrial" entertainment—a product of the same system that created U.S. Steel and Standard Oil.

Furthermore, the huge success of vaudeville, which built enormous
fortunes for a lot of people, was predicated on the expanded middle
class of industrial society. "For the first time in America, a form of
entertainment was developed that offended virtually no one and ap-
pealed to all classes,"[33] says Wilmeth.

Although working people did constitute a substantial proportion of
the audience for vaudeville, the poorest among them could not afford
to go often.[34] There was, moreover, the language barrier for many.
The middle class, which was still living in cities and towns, not in the
suburbs, had certain attributes which made it important to show business. For one thing, middle-class people had greater leisure than laborers in the mills and mines. Clerks and accountants did not routinely
work the sixty-hour week that was typical of the mills and mines, and
furthermore, they were not so exhausted at the end of the day. For
another thing, they had considerably more money to spend on entertainment than laboring people did, and were far better able to afford
to go to the more expensive palaces where the top acts worked. For yet
a third—and this is important—the members of the middle class could
not generally avail themselves of some types of entertainment that were
open to working people. A young male accountant might be permitted
an occasional visit to a bar, and would probably also have a periodic

fling in a vice district; but he could hardly habituate such places without risking his middle-class status; and he certainly could not bring his wife or his fiancée to such places. As for middle-class women, far from being able to go to taverns and the like, for the most part they could not go out at night unless escorted by a male who stood in some clear relationship to them as husband, fiancé, brother, or other relative. Through the years of the vaudeville boom there were arriving other entertainment centers, particularly the new institution of the cabaret, and the big, gaudy restaurant which offered entertainment and, after 1910, dancing. But these places were expensive, and they still appeared slightly tainted to many people of the middle class. Vaudeville was acceptable to all but the very religious; it was relatively inexpensive; and there were vaudeville theaters everywhere. Vaudeville thus gave show business a respectability that it had not always had; it brought it into the American mainstream; it made it seem legitimate in a way that much of the rough entertainment of the early day was not.

The industrial city was critical to the creation of vaudeville, but at the same time it made it *necessary,* or at least inevitable. There simply were fewer things for city-dwellers to do than there were in the country with its streams, lakes, open fields, woods. In a day when the modern programs of city parks, gymnasiums, swimming pools, and the like were only beginning to be developed, there was little to do outside of the home, and for many people, especially in the tenements, little to do inside of the homes, either. Younger readers may have trouble imagining a life without radio, records, television, movies, or public softball, tennis, and basketball courts. For people who could not, or would not, go to saloons and low dives, vaudeville filled an enormous hole in their lives. It is thus hardly surprising it grew to such proportions.

The new vaudeville was paralleled and eventually intertwined by another element in show business aborning, the commercial music industry, or Tin Pan Alley as it very quickly came to be known. The idea of popular song was hardly an invention of the late 19th century. It is probable that the ancient cave painters sang songs and even played on musical instruments;[35] and then as now some songs must have been better liked, and sung more often, than others. The first songbooks in the United States were, like much else, imported from abroad, mainly England, but by the early part of the 18th century music was being published here; and by the end of the century a small, but growing, music publishing business existed in the new nation.[36] Through the 19th century the music publishing business expanded into a substantial industry, with a good deal of money being made by publishers, if not necessarily by the songwriters. Popular song, or "vernacular" song, as it has been termed by H. Wiley Hitchcock, an authority on American music, came in many varieties: spiritual folk songs made of religious lyrics attached to well-known tunes—the song we know as "Go Tell Aunt

Rhody" had both religious and secular lyrics; minstrel and so-called coon songs abstracted from black folk music for general audiences; hymns; sentimental ballads; and much else.[37] Some of these songs, like James A. Bland's "Carry Me Back to Old Virginny," Dan Emmett's "Dixie," and a number of Stephen Foster tunes were immensely popular and have remained part of the American musical tradition to this day. By the last decades of the century, marches, used both as concert pieces and for dancing—especially the twostep—were big sellers. John Philip Sousa, one of the finest composers of American vernacular music, was famous not only for his celebrated band, but for his compositions like "Stars and Stripes Forever."

The 19th-century music publishing industry was relatively genteel, as befit the Victorian culture it was embedded in. The major publishers usually issued a wide variety of music, not only sentimental ballads of the day but choral music, piano solos, complete scores for orchestras. Their wares were listed in catalogues and were usually bought by mail. The industry did not aim for quick hits, which would disappear in a few weeks, but songs that would go on selling for years. The industry, furthermore, was spread across the United States, with major music publishers operating from Cincinnati, St. Louis, and Baltimore, as well as Boston and New York.

With the coming of vaudeville all that changed. According to Charles Hamm, an authority on American popular song, the shift was marked by the establishment of the T.B. Harms Company in New York in 1881, the same year Tony Pastor opened his Fourteenth Street Theater.[38] Harms had a hit very quickly with "Wait Till the Clouds Roll By," which Hamm calls "the first 'Tin Pan Alley' song, since the ensuing success of the Harms company eventually revolutionized the music publishing business and changed the character of American song."[39] Over the next few years Harms had other hits in "When the Robins Nest Again" and "The Letter That Never Came," and very quickly other ambitious men formed companies to emulate Harms's methods. During the late 1880s and 1890s such Tin Pan Alley giants as M. Witmark and Sons, Shapiro, Bernstein and Company, Jerome H. Remick and Company, Leo Feist Music Publishing Company, and others were founded. Now they were all in New York City, mainly clustered around Union Square on East Fourteenth Street, then the heart of the New York theater district. As show business moved up Broadway, Tin Pan Alley followed until most of the publishers' offices were on West Twenty-seventh and Twenty-eighth streets, mainly off Broadway. And this was the location that was named Tin Pan Alley by the successful songwriter Monroe H. Rosenfeld.[40]

The publishers quickly realized that the best—indeed, nearly the only— way to get their songs widely known was to have them performed. "A new song must be sung, played, hummed, and drummed into the ears of the public, not in one city alone, but in every city, town, and village, before it ever becomes popular," said Charles K. Harris, a song publisher and author of the immense hit, "After the Ball."[41] Although performers in saloons, concert gardens, and the like could be helpful, vaudeville was the best medium in which to display new songs. Courtship of vaudeville stars by publishers was intense: offers of drinks, meals, board bills escalated in the competitive scramble to outright bribes, percentages of the song involved, and bonuses paid when the performer returned to New York after a successful tour.

There were huge sums to be made, not the least because the industry was being subsidized by the songwriters: only a very few composers were able to command royalty arrangements; in most cases songs were sold outright for $10 to $25, and even this was not all profit, for if the composer was not able to write his own lyrics, or lift them from the public domain, he had to pay a lyricist perhaps $5 for the words. Hamm says, "A picture comes into focus of a ferociously competitive industry dominated by the publisher, with the singer his most powerful ally. Songwriters seem to have been looked upon, mostly as a necessary evil."[42] As a consequence, the most astute and aggressive songwriters formed their own publishing companies, in order to make something for themselves.

It is clear that the meretricious attitude which is characteristic of the music industry has been with it from the early days: the hit mentality was there long before Tin Pan Alley left Fourteenth Street. The hits were what brought in the big money, and every song that got published had to be a potential best-seller. Publishers worked out formulae based on what they thought had made their top songs successful and expected their composers to obey the rules. Every hit was followed by a spate of imitations. Songs, which had never been terribly complex in any case, were simplified down to pure bone.[43] In an earlier day popular songs frequently told complete, if rather basic stories: "After the Ball" related the tragic story of a man who mistakenly concluded that his girl friend had been unfaithful to him, and only discovered the truth when it was too late. Most songs of the earlier period consisted of verses that advanced the tale, answered by a chorus, usually identical each time. Now the verse dwindled to a type of introduction, usually sung only once, and the narrative lyric disappeared. Songs were meant merely to set a mood or describe a situation, as for example "Take Me Out to the Ball Game" or "Wait 'Til the Sun Shines Nellie." The music, invariably in a major key, depended upon three or four basic chords and patterns of secondary dominants. With such limited harmonic structures it was inevitable that many songs had virtually identical chord progressions.

As Hamm points out, a startling number of these songs were tragic. We think of pieces like "In the Shade of the Old Apple Tree," "My Gal

Sal," and "You Tell Me Your Dream," as cheerful love ballads, but in fact all end with the news that the beloved is dead.[44] A lot of them were woeful tales of the ill that befell country boys and girls moving to the big city, and there were all of those lyrics that dwelt on happy old days back on the farm. By the turn of the century the music business had become huge, with many songs selling over a million copies of sheet music, and some selling more than two million.

Involved in the spectacular rise of the popular music business was a boom in the piano industry.[45] Americans had been producing pianos as early as the 1770s, but it was with the development of the metal frame, which replaced the heavy wood framing needed to support the tension of the strings, that a large-scale American piano industry came into being. The Chickering family, who did much to develop the metal frame and brought other innovations to the piano, were the first to go into large-scale production of the instrument. Later, the Steinway family, German immigrants, began to offer competition to the Chickerings. Particularly in the 1870s and after, the Victorian belief in the uplifting quality of art brought a boom in touring artists, many of them Europeans. Concerts of the piano music of Chopin, Liszt, and others were widely popular.[46] The piano became a major status symbol, for possession of one suggested a degree of cultivation. Furthermore, it was believed that "good" music had a purifying effect, and children trained to play the classics would be less likely to fall into evil ways. By the late 19th century any family who could afford a piano and had the space almost automatically bought one. In 1829 there was one piano sold in the country for every 4800 people; by 1870 it was one for every 1540, by 1890 one for every 874, and in 1910, a peak year, one of every 252 Americans bought a piano.[47]

Thus there were, by the rise of Tin Pan Alley in the last decades of the 19th century and the first decades of the 20th, pianos in millions of American homes, as well as countless numbers in saloons, brothels, beer gardens, and the like. Without these pianos, and people who had been trained to play them, however badly, a large-scale music industry could not have come into existence. Once again the industrial city had triumphed over Victorianism: millions of pianos that had been bought for the performance of Chopin etudes were now being rattled to Tin Pan Alley tunes.

The popular music industry was not so directly a product of the new industrial society as vaudeville, because its audience did not have to be collected in one place; but because it existed only as part of the new forms of entertainment, especially vaudeville, it was indirectly tied to the rising city. Furthermore, like any national industry the web of copper and steel uniting the country was important to its ability to do business rapidly and in large volume, especially as it was headquartered in New York City.

Moreover, a staggering proportion of the people who created it and ran it for decades came out of the big city immigrant slums, or were children and grandchildren of people who had. Music, after all, was something people could enjoy even if they spoke broken English. We have only to look at the names: Remick, Bernstein, Witmark, Shapiro, Cohan, Berlin. Says Hamm, "These publishers were city people, many of them first-generation Americans with little contact with American life and culture outside New York City, and little or no knowledge of the Anglo-American and Afro-American music that was the native musical language of many millions of Americans."[48] It will not do to exaggerate this: many of the people in the music industry were people of the old stock, and a lot of those of immigrant stock were second- and third-generation Americans. Nonetheless, the immigrants played a major role in the new music business. Inevitably, they were pouring their own attitudes, ideas, and concerns into the musical forms they found in the United States. In the course of this they changed popular song. Hamm says, "Tin Pan Alley did not draw on traditional music—it created traditional music."[49]

In part because of their lesser musical training—George M. Cohan could play only four chords on the piano,[50] and Irving Berlin had to hire somebody to help him harmonize his songs[51]—they simplified their songs both in words and music. They tended to focus much more heavily on male-female relationships: thirteen out of sixteen of the songs on Hamm's list of major hits from 1892 to 1905 are essentially love songs.[52] The old narrative or descriptive song was disappearing. The popular song had become *personal*—a statement of something the singer felt strongly about, usually someone of the opposite sex. This was hardly surprising, given the immigrants' tendency to be more directly emotional in expression than was the case with people from the old Anglo-American culture. The personal song was not, of course, a new invention: "All up and down the whole creation, sadly I roam," was the first line from one of the most popular of all songs of the earlier day. But hereafter, the personal song would become the dominant mode, the foundation on which the popular song of the 20th century was built. The great tunes of the Golden Age were endlessly cut to this pattern: "Star Dust," "Dancing in the Dark," "Embraceable You"—the list is endless. And even long after the great years were over, the personal song continued to be at the center of popular music, as witness "I Can't Get No Satisfaction," "Love Me, Do," and thousands more. Needless to say, those were about the needs of the self.

The new Tin Pan Alley song was quickly adopted by the middle class. Exposed to it through vaudeville, these people found it attractive, perhaps simply for its personal, expressive qualities. The middle class was a very important segment of the market for Tin Pan Alley; once again middle-class families were more likely to have the pianos, the money, and the leisure on which consumption of popular music depended. In time the phonograph record, not sheet music, would become the main medium for the popular song; but at the beginning of Tin Pan Alley the money came from sheet music. The majority of immigrant families lacked the space for a piano, the leisure for practice, the money for anything. Some of course did not, especially those who came from the old immigration of the Irish and Germans, who by 1900 had been in America for two or three generations; but at the beginning, Tin Pan Alley needed a middle-class audience. And it got it.

Related to the emergence of the commercial music industry in the 1890s was the appearance of a group of quintessentially American musical forms, which would heavily inflect, indeed to a considerable degree shape, popular music in the United States and eventually the world. These were ragtime, the blues, jazz, and the so-called syncopated songs. They were related, in that all of them were produced at the crack where black and white cultures in the United States were grinding against each other like tectonic plates. All were, in form and basic harmony, European; all were rhythmically, and perhaps in spirit, drawn from the black culture.[53]

The development of these forms is complicated and not well understood. To put it briefly, the black slaves carried to the United States their rhythmically complex African music, and in some cases even their instruments. Through the syncretic process the black slaves began adapting the musical forms they found in the United States to express the content of their own music, and in time there evolved a black folk music that was neither African nor European but a thing of its own.[54]

Whites, at first in the South, and then in the North, came to know this music, and many saw its worth. Heavily Europeanized, it became the basic "plantation" music of the minstrel shows, utterly familiar to Americans through the hits of Foster, James Bland, and others. In the post-Civil War period a few northerners went South to study black music, and some transcriptions of it were published.[55] Finally, in the 1870s a student choral group from Fisk University, called the Fisk Jubilee Singers, traveling to raise money for the college, was enormously successful singing "spirituals," black religious music again Europeanized for the benefit of Americans accustomed to a different sound.[56]

This music had certain recognizable characteristics—a tendency toward plagal rather than perfect cadences, and the consequent frequent use of the sixth degree of the scale; and rhythmically, a good deal of syncopation, especially in the employment of the figure perhaps best known as the first three notes of "Turkey in the Straw." Through the 19th century thousands of plantation songs and spirituals were published, and in time the form came as natural to the American ear as the church hymns that had been a musical staple for generations.

With the movement of blacks into minstrelsy, many of them began composing in this style, and eventually there emerged a variation of it which came to be called the "syncopated song." Between about 1890 and 1915 a group of black poets, composers, and performers, many of them well schooled, created for the Broadway theater in New York about thirty musical shows, usually built around black themes. The plots tended to be rather sketchy, and served mainly as platforms for song and dance. These shows were produced and performed almost entirely by blacks; surprisingly, they frequently played to semi-integrated audiences, with blacks seated on one side of the house, whites on the other. Among the creators of these shows were Will Marion Cook, J. Rosamond Johnson, Paul Dunbar, and James Weldon Johnson. Many of them featured Ernest Hogan and the team of George Walker and Bert Williams, the last being the best-known black performer of the early part of the 20th century.[57]

These black shows certainly contributed to the development of the musical; and they produced a body of highly popular songs. Thomas Lawrence Riis, in his study of these shows, has noted that four of twenty-four best-sellers in 1902 were written by blacks, and that in 1901 six of the seventeen songs that sold over 100,000 copies were also by blacks.[58]

The syncopated songs, like the spirituals and plantation melodies, also helped to acclimate the American ear to black musical forms. Ragtime, thus, came as no great surprise. The genesis of the rags is lost, but it is reasonable to suspect that it grew out of an attempt to transfer the minstrel banjo style to the piano. Dan Emmett's "Pea Patch Jig," a banjo tune adapted from black plantation music well before the Civil War, looks strikingly like ragtime.[59]

In any case, there was apparently some sort of piano ragtime being played by the 1880s, mainly by self-taught black pianists drawing on elements taken from a number of sources. St. Louis and the Missouri railhead, Sedalia, where there was a famous vice district, were particularly important loci of ragtime: Scott Joplin's classic "Maple Leaf Rag" was named for a black-and-tan saloon in the Sedalia vice district.

Through the 1880s and 1890s ragtime crept into the American consciousness, primarily through exposure in saloons, black and tans, and the vice districts. By the later years of the 1890s they were being issued by Tin Pan Alley publishers. In 1899 Joplin published both his "Original Rags," and "Maple Leaf Rag," which rapidly became the most celebrated piece in the genre. By 1901, according to Edward A. Berlin in his standard work on the subject, most of ragtime's "basic conventions were well established."[60]

Ragtime went on very quickly to become a major component of popular music. It was speeded along by the first form of mechanical entertainment to become widely used, the player piano. This was an ordinary piano to which was attached a device which activated the keys, and it made it possible for people who had nobody in the family competent to play ragtime to enjoy the music. According to one account, "Between 1895 and 1912, there were more player pianos in the United States than bath tubs."[61] Even as late as 1923 some 205,000 of the 348,000 pianos sold were players.[62] The player piano was crucial to the popularity of the more difficult rags, like the later pieces of Joplin, which gave even fairly competent amateurs trouble.

However, ragtime was best known to Americans in the form of the "ragtime song," which was produced by simplifying the most popular strain from a rag and fitting it out with lyrics. Eventually ragtime songs were written apart from any piano compostion: Irving Berlin's immense hit, "Alexander's Ragtime Band," was one such.

For a period the United States was ragtime mad. Sheet music, piano rolls, and eventually recordings of rags sold in the millions. But by 1920 it was being pushed aside by forms which seemed more up-to-date: jazz, the blues, and especially jazz-inflected dance music of Paul Whiteman and others. Nonetheless, for the student of contemporary culture ragtime has great significance, for a direct line can be traced from it through jazz to rhythm and blues and thence to rock.

Jazz, which would give its name to the decade of the 1920s, evolved out of ragtime and the blues in New Orleans and its environs just after the turn of the century. It was first played by small combinations of horn and rhythm players who generally worked without music, and perforce improvised a portion of the music. Contrary to the legend, it spread first to the West Coast in about 1910, and then began to move into other northern cities, principally Chicago. Although some of the early jazz bands worked in vaudeville and in taxi dance halls, much of it was played in the saloons and dance halls of the vice district, such as Freiberg's, the main institution in the Levee, and Purcell's on the Barbary Coast.[63]

By 1916 it was fairly well-known to habitués of low haunts; and early in 1917 Reisenweber's, a famous New York City "lobster palace," brought a group of white jazz musicians, called the Original Dixieland Jass Band, to one of its rooms. The group was a sensation, and its Victor records quickly became enormous hits. By 1920 the music had a wide following in the United States and was beginning to be taken up by intellectuals as a significant new form of art.

The blues followed a similar pattern of development. Although efforts have been made to place its origins well back into the 19th century, there is little evidence for the blues before 1900. It is probable, but far from certain, that they evolved from work songs in the so-called Delta area of northwest Mississippi, a cotton and soybean region, around the turn of the century. They were heard in the first years of the 20th century by Ma Rainey, later termed "the mother of the blues,"[64] and by W. C. Handy, the most famous composer of formal blues numbers.[65] As sung in forms suitable for vaudeville by Rainey, Bessie Smith, and other black female vocalists, they quickly grew popular among both rural and big city blacks who heard them by way of traveling tent shows. Then, in 1912 Spencer Williams's "Dallas Blues" was published, followed by W. C. Handy's "Memphis Blues." In 1914 Handy wrote his famous "St. Louis Blues." Interest grew. In 1920 another black singer, Mamie Smith, made a recording of "Crazy Blues," which became a surprise hit and triggered a vogue for the blues which would last through the 1920s.

These forms, derived from black musical traditions, eased themselves into the American mind by stages. First came the plantation songs with their slightly exotic plagal harmonies and modest syncopation; then came the only moderately more complex syncopated songs; then the somewhat more thorny ragtime; then the rough, densely polyphonic dixieland jazz; and only finally the classic blues, with their non-diatonic tones and ambiguous rhythms that eluded transcription.

These forms were mainly products of the industrial cities. Ragtime came out of the saloons and vice districts of St. Louis, Sedalia, and elsewhere. Jazz was born in New Orleans, especially in the black honky-tonks, and was brought into daylight through the vice districts of San Francisco, Los Angeles, Chicago, and other cities. The blues, although rural in origin, became popular only after they were brought into the cities: it is not surprising that some of the first big blues hits were named after Dallas, Memphis, and St. Louis. And of course they achieved nationwide acceptance only when they had been taken up by the music industry as novelties that could be sold to the public as the newest thing. It is simply true that without the industrial city there would have been no blues, no jazz, and all that flowed from them. It is no accident that ragtime reached its peak in 1909–11, that the first blues sheet music was published in 1912, that jazz was first mentioned in print in 1913: this was the moment when the new industrial city was firmly in place.

And once again, just as Tony Pastor and his imitators had set out deliberately to create an entertainment medium which would be acceptable to the middle class, so did musical entrepreneurs attempt to devise forms of these new musics that could go into middle-class homes. Scott Joplin, who was trained in classical composition, was deliberately using in his rags ideas drawn from serious music; eventually he wrote both a ballet and an opera, *Treemonisha,* which contained some pieces drawn from black forms as well as a good deal of straightforward theater music. Paul Whiteman, with considerable assistance from Ferde Grofé, a trained musician who was the son of immigrant parents, set out to produce a refined version of the rough polyphonic jazz of the honky-tonks and dives, which was ultimately called "symphonic jazz." W. C. Handy, in writing down the rough blues he picked up in his travels, brought them into the diatonic system, in order to make it possible for mainstream Americans to comprehend them. And Irving Berlin and others were reducing fairly complicated ragtime to the simple ragtime songs.

What we begin to see, then, is that the entertainment industry which would come to dominate the lives of Americans was in large measure created by modifying forms which had originally been devised for the amusement of working people, especially immigrant and black working people, in their saloons, dance halls, and honky-tonks, to suit middle-class tastes.

It was, then, a compromise; the middle class spawned by the industrial city needed entertainment, and it was finding something that interested it in working-class saloons, and especially in the vice districts: Paul Whiteman and Ferde Grofé, who devised the formula for the modern dance band, first learned about jazz on the Barbary Coast; the manager of Reisenweber's found the Original Dixieland Jass Band in "a dive";[66] middle-class whites frequently went to what were called midnight rambles—black shows put on for white audiences—to hear Bessie Smith and other great performers, and similar midnight rambles were given in Lincoln Gardens, a black dance hall in Chicago for the benefit of whites, many of them middle class, who were interested in the new jazz music.[67]

But the middle class, still clinging to the Victorian ethic into the 1920s, did not want working-class entertainment, with its large erotic component, unalloyed; it would have to be brought closer to the forms and mores that middle-class people were at home with. And people who were able to work this trick, like Whiteman, Gershwin, Handy, and hundreds of others, got rich.

CHAPTER 8

The Mechanization of Entertainment

The entertainment explosion which took place in the latter years of the 19th century was driven to a considerable degree by the itch for invention that characterized the age. The same kind of people who were shackling electricity for human uses and developing steamships and automobiles were also inventing devices whose main function was to amuse. And they were doing it for the same reason: to make money. Thomas Alva Edison, who worked out the first sound recording machine and contributed to the development of the motion picture, at first pursued neither idea because he did not see how money could be made from them.[1] Others did, however; and in the end it turned out that a great deal of money could be extracted through these new systems for amusing people.

Almost from the moment that it became clear, in the 1830s, that it was possible to make photographic pictures, people began to wonder if there was not some way to make pictures that would move.

A number of people contributed to the development of moving pictures, but credit for working out the basic system is generally given to a man who chose to call himself Eadweard Muybridge, who started by making serial pictures of animals in motion. In 1888 the ubiquitous Thomas Alva Edison put one of his employees, William Kennedy Laurie Dickson, to work on the problem.[2] By 1889 Dickson had developed "a fully functioning movie camera, which Edison termed the *kineto-graph,* at the same time claiming its invention," according to John L. Fell, an authority on early film.[3] Edison, however, like most inventors of the day, was an eminently practical man who thought in terms of the commercial possibilities in a new technology, and he could not see how money was to be made from what appeared to be basically a toy.

Others, however, were more percipient. By the 1890s dozens of very brief films had been made. Many of them drew upon vaudeville, and showed Sandow the strong man flexing his muscles, and Annie Oakley doing her trick shooting routines. The potential for sexual display was not missed: some of the earliest films showed cootch dancers, windy days with skirts flying, husbands stealing kisses from servants and typists.[4] These first movies were exhibited in two ways: by means of automated coin slot machines—so-called peep shows—which allowed the viewer to watch a brief movie through a lens; and by the projection of a magnified image on a screen. Eventually Edison woke to the fact that there was money to be made and he joined with another inventor, Thomas Armat, to manufacture a film projector Armat had developed; they called it the Vitascope. It was introduced at Koster and Bial's music hall in New York in 1896 to a good deal of acclaim, and it quickly became clear that the movies had a future.[5] Entrepreneurs rushed in from every side, some to make movies, some to appear in them, some to show them and exploit them in other ways.

At first most movies were shown as acts on vaudeville bills, for their novelty value: many of them consisted simply of shots of ocean waves, dancers, a train rushing towards the audience. However, as it became clear that audiences were fascinated by the new medium, promoters began to open theaters intended solely for the showing of movies. Little movie theaters existed from as early as 1896,[6] but the real boom in them took place in 1905 and after. The usual admission fee was a nickel, which led to the name nickelodeon. By 1907 there were 2,500 of them, mostly in the nation's big cities, and within two or three years there may have been 7000 to 10,000.[7] In 1907 there were an estimated two million paid movie admissions a day; in 1911, five million; in 1914 seven million.[8]

The explosion of movie theaters was made possible by the ease with which they could be started. All you needed, besides a low-rent store front, was a couple of hundred kitchen chairs, a sheet, a projection machine that cost $165, a piano, which could be rented, and odds and ends of lighting fixtures. One authority has calculated that a store-front movie theater could be opened for $600.[9] By 1910 or so many operators owned as many as a dozen theaters. The nickelodeons, it quickly became clear, were money machines; some of the more astute operators found themselves earning profits of $25,000 a year, which at the time was enough to allow a man to account himself wealthy.

In the main these early movie houses showed one-reel films, which ran for ten or fifteen minutes. Some of them were travelogues, prize-fights, re-creations of famous disasters, like train wrecks. But very quickly the story, however crude and brief, became the staple of the movies. Fell quotes one nickelodeon operator as saying in 1907, "The public wants a story. We run to comics generally; they seem to take best . . . when we started we used to give just flashes—an engine chasing a fire, a base runner sliding home, a charge of cavalry. Now, for instance, if we want to work in a horse race, it has to be a scene in the life of the jockey, who is the hero of the piece—we've got to give them a story; they won't take anything else—a story with plenty of action."[10] Says Fell, "Film fiction resolved itself into genres identical with earlier popular entertainment: dime novels, boys' books, and stage melodramas."[11] Right from the beginning there would be horror films, westerns, romances, adventures.

At the beginning virtually all films shown in nickelodeons were one-reelers: D. W. Griffith, the medium's first genius, made hundreds of them for Biograph. But Griffith was a visionary who was eager to develop his craft, and by 1910 he was pushing Biograph to let him make two-reelers and even longer films. These proved successful with the public, and following Griffith's lead, the industry began to develop what came to be called the feature film. The movement was helped along by the example of imports like *Quo Vadis*, which was made in Italy in 1913 and was eight reels long.[12]

As had been the case in vaudeville, at first the players were relatively anonymous; but Griffith began to bring into the movies the billing system that had proven important to vaudeville, and very quickly the movie star was born. Under Griffith's guidance Mary Pickford, Mae Marsh, and the Gish sisters became stars, earning enormous sums for the period: by 1916 Pickford could make $10,000 a week, and the cost of making a single film was rising to $100,000.[13] As the film boom grew through the 1920s money exploded and was flying through the air everywhere.

There has been some debate among students of early films about who made up the first audiences. It has generally been believed that they consisted largely of people from the immigrant slums. The point was that you did not need to speak English to be able to enjoy most silent films; and even after subtitling became common, there was usually somebody in the audience who could translate. In these early days the reputation of the movies, and movie houses, was bad: the middle class, especially women, were "reluctant to peer inside of the early movie houses," says Russell Merritt in his study of the subject.[14] This was in part because of the sexual element that was present in some early films; in part because middle-class people did not like associating with uncouth immigrant audiences who ate, spat, smelled, shouted, and gesticulated; and, more broadly, because of the Victorian attitude that did not approve of activities which were not uplifting or socially useful.[15] The middle-class view was expressed by a Chicago judge who, in 1907, announced that the nickelodeons "cause, indirectly or directly, more juvenile crime than all other causes combined."[16]

But, just as the vaudevillians had sought middle-class audiences, so did the new movie industry quickly come to see that it needed the middle class if it was to grow into something substantial. In particular, they wanted "the New American woman and her children." Says Merritt, "The seduction of the affluent occurred . . . between 1905 and 1912, in precisely that theater supposedly reserved for the blue-collar workers."[17] Once again the idea was that if you could make the medium acceptable to women, the whole middle class would come. Exhibitors began offering women and children half-fare or even free tickets, controlling rowdiness in the theaters, and cleaning them up generally.[18] Producers cooperated by making films aimed at women. "'Female protagonists far outnumber males, dauntless whether combating New York gangsters, savage Indians, oversized mashers, or 'the other woman.' "[19]

It is particularly important to note that film producers began consciously to adopt for films "the outdated moral code of the Victorian era that required vice to be punished and virtue rewarded . . ."[20] It was a question, once again, of those "influences." This was, we remember, the time of the Victorian backlash, a period when the middle class was filled with exceedingly ambivalent feelings about the new industrial world aborning. But this ambivalence left the film producers caught in something of a dilemma, for what attracted movie audiences were exactly those suggestive elements that Victorians feared: sexuality, crime, violence, pleasure. For decades thereafter the movie industry had constantly to tread a fine line between too much sex and violence and too little. It was always a matter of so far and no farther; but with the amount of money involved, producers were willing to make the effort.

This was particularly so because, right from the beginning, the young, subject to those influences, have been a very significant portion of the movie audience. A 1914 study of Ipswich, Massachusetts, showed 69 percent of boys in grades five through eight went to the movies once a week or more; in high school it was 81 percent.[21]

By 1912 movie operators were moving out of the storefronts into converted vaudeville houses, and raising prices. These were precisely the theaters the middle class was accustomed to going to. Newspapers began giving the medium increased coverage. As ealy as 1910 Horace M. Kallen published an "esthetic of film" in the *Harvard Monthly;* in 1911 *Motion Picture Magazine* was started;[22] and in 1914 President Woodrow Wilson was persuaded to watch a film with his family and members of his cabinet.[23] And by that moment the issue was settled, and "the producers had widened the audience to include the middle class."[24]

Like vaudeville, the movies were a product of the industrial city. The medium needed a large, compact population to draw from, good local transportation, a new technology, and at first the immigrant mass in need of cheap amusement. As had been the case with variety, jazz, and popular music in general, a form devised to suit the tastes of working people was modified to make it acceptable to the middle-class ethic, with its residual Victorianism.

The phonograph, however, was an exception to this general rule, for right from the beginning it belonged to the middle class. The first talking machine was worked out by Edison. He saw the instrument primarily as a tool for businessmen, who could use it for dictation, to record telephone conversations, and for various educational purposes.[25] He was scornful of the idea that it might become a toy for entertainment purposes; like most men of his time he saw life as a serious business not to be treated as a lark.

For a considerable number of years, however, it appeared that the instrument would be of not much use. Reproduction was poor, the cylinders then in use lasted for about two minutes, the apparatus was too expensive for most families to purchase. But through the 1880s various improvements were made, and by 1890 or so the public was beginning to pay attention. Despite Edison's opinion, the machines were sold primarily to saloons and drugstores, where they were used for amusement as early versions of the jukebox: the auditor deposited a nickel in a slot and listened through earphones. One New Orleans drugstore claimed to be making $500 a month from its talking machine, but more typically they would bring in $50 a week.[26] This calculates to something like 150 playings a day per machine, which suggests that even in this early stage Americans were listening to records at least a million times a day.

Through the 1890s there were further improvements. The most significant of these was the development of a practical disc recording. The disc had been patented by a German immigrant named Emile Berliner in 1887 but not marketed until 1894.[27] It quickly overtook the cylinder, and by the turn of the century had come to dominate the field, although cylinders were produced for some time thereafter.

These early recordings featured brass bands, cornet solos, banjo solos and duets, comic monologues, whistlers, and the like. They were surprisingly good, considering the primitive equipment used to make them. The main weakness was that the artist had to deliver his sound into a huge conical horn, and this made it difficult to effectively record large groups, like symphony orchestras, although a considerable number of marching bands were recorded fairly well. The human voice, however, recorded very well, and even when electrical recording was introduced in the mid-1920s there were some record fanciers who thought that the quality of sound produced by the electrical method was inferior to that produced by the old acoustical system.[28]

Through the 1890s sales were very slow, due mainly to the fact that the machines were still too expensive for ordinary incomes. But in 1901 Eldridge Johnson, who had made important improvements to the sound quality of discs, established Victor to make both records and machines.[29] Prices began to fall, and sales began to boom. Nonetheless, prices for records, to say nothing of the machines, were still too high to permit working people to buy them. The consequence was that the early records were aimed at the middle class. And it is therefore not surprising that the first recording star was the Italian singer, Enrico Caruso. Roland Gelatt, in his study of the industry, says, "It is generally agreed today that Caruso's Milan series of March 1902 were the first completely satisfactory gramophone records to be made."[30] Needless to say opera was to the Victorian taste—uplifting and spiritual—and a great many opera singers were recorded in this early period. The phonograph made Caruso wealthy: by his death in 1921 he had earned $2 million from phonograph records alone[31]—the first of a long line of performers who would earn astonishing sums from recordings.

From this moment it was all upwards and onwards: Victor grew from assets of $2.7 million in 1902 to $33.2 million in 1917.[32] After 1910 or so the basic patents began to expire, and by 1914 other companies were coming in to challenge the dominance of Victor and its chief competitor, Columbia. The companies were recording everybody and everything: comics, dozens of opera singers hoping to capture a portion of Caruso's audience, massive amounts of popular music, ragtime, stump speeches, and after the success of the Original Dixieland Jass Band's records in 1917, increasing amounts of jazz.

But more and more through these years the staple of the recording industry became dance music. Many early records were specifically labeled "one-step," "two-step," "fox-trot," and the like. Everywhere around the United States people were rolling up rugs after dinner and dancing to records. It was a fad, a vogue, a craze. By the 1920s, furthermore, the price of the machines had finally gotten down into a range that working people could afford. Records were still expensive: seventy-five cents was the standard price, which took a laboring man two or three hours to earn. But so strong was the lure of music that even poor blacks living in ramshackle houses in city slums, like Harlem and Chicago's South Side, were able to buy the new recordings of their favorites, like the great blues singer Bessie Smith, or the hot trumpeter Louis Armstrong.[33] In this instance the movement from working class to middle class was reversed: while Victor continued to sell its famous Red Seal classical records to the better educated, by the 1920s working people were being catered to: the record catalogues of the time are divided into sections devoted to Irish, Polish, Italian, German, and other ethnic music, as well of course of the "Negro" music.[34]

The recording industry was not so directly dependent on the industrial city for its development as vaudeville and film were. People did not have to be gathered in one spot to buy the commodity. Nonetheless, by the 1920s the recording industry was becoming enmeshed with

Tin Pan Alley and the popular music business in general. With vaudeville on the wane, the record was now the best way to get a song heard. Song publishers used whatever means they could to induce big name artists to record their tunes; at times they even paid for recording sessions themselves: hiring a band, paying for arrangements, and then offering the recording to the company gratis.[35]

Thus, the record business willy-nilly became part and parcel of the blooming new entertainment industry which the industrial city had thrown up. It could have existed apart from the cities; but in fact it did not.

Running closely parallel to the rise of vaudeville out of saloon entertainment was another branch of the entertainment industry which has proven to have greater staying power: what we have come to call sports. Modern games were not invented by Americans: most of the games we spend so much time watching and much less time playing were invented by the English in the late 18th and early 19th century.[36] Other contributions were made by the Germans in the early 19th century, especially in the area of gymnastics and field sports, with some contribution from the Swedes.[37] The Canadians developed ice hockey; lacrosse was an American Indian game; and basketball was invented in Springfield, Massachusetts. But it was the English who developed, if they did not invent, the ball games which have figured so largely in the lives of people in the 20th century—tennis, golf, the bat games, the kicking games.

It is significant that one of the most important figures in the development of modern sports was Thomas Arnold, headmaster of Rugby, who was attempting early in the 19th century to reform the English public schools, which had become to an extent debauched in the 18th century.[38] Modern sports, then, were initially a product of the Victorian mentality. For these Victorians, just as the main function of art was not to provide pleasure but to uplift, so the function of sport would not be to amuse but to improve. As people must cultivate their minds and elevate their souls, so they must improve their bodies. People, so the Victorians came to believe, had a duty to keep fit. One ought to play at sports: going out for the team or taking an arduous daily swim came to be seen as a virtue.

This was equally true of the Germans, who were developing the idea of physical training: they saw in gymnastics a spiritual element. From the Victorian point of view it *really did not* matter who won or lost, for the point of the contest was to improve the players both physically and morally; what counted therefore was fair play, honor, and a good physical challenge. Even after sports began to be professionalized in the second half of the nineteenth century, the English insisted on maintaining the distinction between the gentleman amateur, who played for the values in the game itself, and the professional, who was in part an entertainer and whose rewards depended upon winning.[39] A residue of this Victorian attitude continues today in the idea that sports are meant to "build character," which is still given lip service in at least some American colleges and universities. For the Victorians it was a real ideal: sports taught fortitude, loyalty, team spirit, a willingness to endure the hard moments uncomplainingly. It was a time when tennis players would applaud each other's fine shots, and would no more think of throwing a racquet or cursing an umpire than they would of belching at the tea table.

The United States was somewhat behind in adopting the new Victorian sports culture. Americans had, of course, followed the lead of the Europeans, especially the English, in taking up horse racing and the blood sports: not only the gentry, but ordinary farmers liked to bet on horses at backwoods tracks attended by rough country crowds.[40]

But the United States, especially New England, carried into the 19th century a residue of Calvinism that told them that idleness was sin, and play the devil's work.[41] Equally sinful was wagering; and the drunken blood-thirst of the spectators around the cockpit could hardly have been attractive to the religious temper.

Gradually, however, the new Victorian view of sport as having moral and character-building qualities drifted across the Atlantic, and the old objections began to dissipate. The 19th century, says one writer, "was a period of beginnings characterized by the gradual breakdown of traditional prejudices against play and amusements."[42]

The first game to benefit from the new acceptance of sport was what came to be called baseball. It began to develop out of its English predecessors in the 1830s. By 1845 there were written rules, and at the same moment Americans did what they would characteristically do— they professionalized the game.[43] By the time of the Civil War there had come into existence leagues, player organizations, newsletters, committees to coordinate rule changes, and the like. The Civil War itself gave the sport further impetus, as bored soldiers in army encampments got up games to while away empty time, in the process teaching the game to others who did not know it. The baseball game at a Civil War army camp was a common sight. And very quickly after the close of the war baseball became a modern spectator sport with a nationwide following.

Richard D. Mandell, a sports historian, says, "The establishment of 'leagues' under profit-oriented managerial control in the 1870s provided the models for other American sports that later professionalized."[44]

Thereafter it all came in a rush. In the 1870s American colleges began holding track and field events in imitation of the English university competitions.[45] Parimutuel betting was developed in the 1860s and

1870s.[46] The National League in baseball was formed in 1876.[47] Boat racing became popular in the 1870s and 1880s.[48] The first meeting of the League of American Wheelmen came in 1880,[49] the first national tournament of the United States Lawn Tennis Association was held in 1881.[50] Boxing, especially after the arrival of John L. Sullivan as the sport's first great champion, began to draw large audiences. "Western Union paid 50 operators to send out 208,000 words of description following John L.'s fight with Jack Kilrain in New Orleans in 1889. When Jim Corbett beat Sullivan in 1892, 300 saloons and billiard halls in New York alone were supposed to have received the news."[51] College football—developed out of the rugby game invented at Thomas Arnold's school a half-century earlier—began to attract a large following in the 1880s, and by the 1890s people were referring to a sports craze on American college campuses, with football the dominant game. Symptomatic of the enormous interest in the game was the fact that in 1903 Harvard, with a student body of some 5000, saw fit to provide a concrete stadium which could hold 57,000 people.[52]

And so it went. The United States Golf Association was formed in 1894.[53] The first international track meet was held at Manhattan Field in New York in 1895. The contestants were the New York Athletic Club and the London Athletic Club, and the Americans won all eleven events. Basketball was invented by James A. Naismith in 1891;[54] and by the turn of the century the sport had become a major element in the American culture.

This exploding interest in sports could hardly be ignored by the media. According to the pioneering sports historian John R. Betts, "Sports had merged into such a popular topic of conversation that newspapers rapidly expanded their coverage in the 1880s and 1890s, relying in great part on messages sent over the [telegraph] lines from distant points."[55] Sporting papers proliferated, and William Randolph Hearst developed for his papers the sports section in the last years of the century.[56] By 1900 baseball players could earn $2,000 annually, as compared with a working man's salary of $700; and by 1910 top players could make $10,000 a year, a very large sum for the time.[57] Says Mandell: "By the turn of the century American sport had evolved into a pattern or system that was unique. Sports spectatorship and (far less) sports participation, sports business, and sports myth were smoothly integrated into American life. The process had been swift, but it had been natural and had gone much farther than the evolution of sport anywhere else in the world."[58]

Sport inevitably cut across all classes; but there was some tendency, through the 19th century in any case, for it to be a gentleman's activity, at least in certain areas. For one thing, developing a skill at throwing a ball or hitting one with something took practice, and practice required leisure time, which neither the farmers of the earlier part of the century nor the immigrants of the latter part had in much abundance. Certain sports, like sculling, sailboat racing, even bicycling, cost money; others, like tennis and golf, required grounds which needed a lot of upkeep. And only the rich could pursue sports like yachting, automobile racing, and equestrianism. The colleges provided the elite who attended them with leisure time for practice, incentive in the form of status which accrued to successful athletes, and the necessary playing grounds and equipment. Football was, basically, college football; and in the first years of this century fully a quarter of major league baseball players were college graduates, at a time when less than 5 percent of the population had college degrees.[59] Other events, like sculling, swimming, and the track and field sports, were also developed at colleges and are even today dominated by college athletes.

But sport was too attractive to be ignored by the majority of the population, and as it was increasingly professionalized it became a route along which working people could escape from the slums. Most games did not require the participants to speak English very well, or the grasp of American customs and traditions. All you had to know was how the game was played; and if you were good at playing it you could be rewarded. (Blacks, of course, were generally disbarred from joining white leagues and teams, although there were some exceptions, most notably in boxing.) Whatever feeling gentlemen from the colleges felt about allowing the lower orders in, the lower orders came anyway. By 1910 or so over half of the baseball players were of Irish or German extraction, an over-representation.[60] Boxing, whose first famous hero was the Irishman John D. Sullivan, drew the attention of immigrants, and when the black Jack Johnson won the heavyweight championship, he drew his ethnic fellows to the sport. During World War I the services often used boxing as a training device; through their wartime service men of all classes became familiar with the sport, which helped to give it more general acceptance; now the middle class could enjoy it. Class lines were being crossed in both directions; it is probably safe to say that in this respect sport has become, especially since the admission of blacks to the main leagues, one of the most democratic aspects of American culture.

It hardly needs to be said that modern sports, like vaudeville, were a product of the industrial city. Like vaudeville, sports needed a mass audience within easy traveling distance of the playing fields. It needed mass transit systems, and a lot of sports grounds were established at the ends of the trolley lines which were then reaching out to city limits. Financiers backing the electrical streetcars at times actually invested in the building of baseball parks at the end of the streetcar lines during the 1880s and 1890s.[61] John R. Betts says, "At the turn of the century the popular interest in athletic games in thousands of towns and cities was stimulated to a high degree by the extension of rapid transit systems."[62]

Sports also needed railroad lines to carry teams from city to city: most leagues were built around intercity, not intracity, rivalries: it was New York against Boston or Chicago, not the East Side against the West Side. When two teams existed in one city, as they often did in the biggest cities, they were usually distributed into different leagues, and did not compete directly. In addition, railways were needed to bring spectators, often in the tens of thousands, to isolated events, like championship fights, boat races, national track meets. The telegraph, and later the telephone, helped to popularize sports by providing results instantly: in small towns around the country newspaper offices often posted inning by inning scores of important baseball games in their windows. Says Betts, "By 1900 sport had attained an unprecedented prominence in the daily life of millions of Americans and this remarkable development had been achieved in part through the steamboat, the railroad, the telegraph, the penny press, the electric light, the streetcar, the camera, the bicycle, the automobile, and the mass production of sporting goods."[63] Modern sport was tight in the embrace of modern technology.

Abetting the rise of sports was the belief, widely held by reformers in the Progressive period, that a good deal of the disorder in the slums was the direct result of a lack of recreation for young people.[64] One of the great functions of the settlement houses was to provide decent occupations for the young, and to this end they formed bands and orchestras, drama groups, clubs and classes of all kinds. Sports were seen as ideal in this respect, for not only did it occupy spare time that might otherwise have been given over to shoplifting or sexual experimentation, but it expended a lot of the restless energy which, so it was believed, often drove the young into unsavory activities for lack of anything else to do with it. Basketball was invented in 1891 precisely to provide a physically active indoor game which could be played in bad weather, especially in the northern cities with their hard winters. All through the period playgrounds, softball fields, running tracks, were built with public funds.[65] "Indeed, among the masses of Americans," Mandell says, "sports came to be considered a civic obligation . . . The Young Men's Christian Association (YMCA) led the way in proposing organized training and team games as methods for absorbing the idle time of poor city boys and instilling in them habits of good hygiene, self-discipline and respect for officials . . . Urban settlement houses and eventually churches also promoted the standard American sports because they presumably developed leadership and built character."[66] It was not just boys, however; girls' basketball, track, and other teams became common even in small towns. The vast system of school sports which we now take for granted was a product of this attitude.

When we look back over the huge entertainment industry that arose in the United States in parallel with the industrial city, we are impressed at how it all came along together. Vaudeville, sports, the commercial music industry began to grow in the 1880s, and were well established in the culture by the 1890s. The movie and recording industry lagged a decade or so behind only because it took entrepreneurs a little time to realize their inherent entertainment value, and then to develop the necessary technology. Sound recording and moving pictures did not come along when they did by chance: they were developed because inventors saw that money could be made from them, and went in search of the technology.

It happened terribly fast. In 1880 there was no recorded sound, no moving pictures, no football, no basketball, no vaudeville, no ragtime, jazz or blues, no Tin Pan Alley. By the first decade of the 20th century all were in place and were—or were about to be—matters of intense national concern. It does not take much imagination to grasp the sense of dislocation millions of Americans, born in the 1870s and 1880s to a world in which "fun" meant sledding, word games, picnics, on finding themselves as young adults simply deluged with professional diversion in mammoth cities where everything seemed to be run by electricity.

What, finally, was the meaning for Americans—and ultimately for the world—of the sudden appearance of this huge system of mass entertainment, with its vast machinery for bringing music, games, shows of all kinds to the American public?

Today we so take for granted our deep involvement in entertainment that it is difficult for us to remember that it is actually a great novelty in human experience. Up until the last decades of the 19th century the vast majority of human beings, stretching back into prehistory, rarely had the experience of being professionally entertained. As hard as it is to believe, the bulk of humanity until recent times never saw a "show" of any kind.

Instead of being entertained, people everywhere amused themselves. They did so almost always actively, and in groups. They danced, they sang, they competed in games, they gambled at cards and dice, they sat in front of fireplaces and around taverns swapping stories, playing word games, posing each other conundrums. Even as late as the 1950s, and perhaps later, boys shot marbles in the spring, flipped baseball cards, played jackknife games like mumblety-peg passed down from generaton to generation. And up to this moment girls in big cities, suburbs, and small towns play traditional sidewalk games like hopscotch and jump rope.

In my view there is a significant, qualitative difference between playing a game and watching it being played; between making music and listening to it; between telling a story and watching one being acted out. The one is active, the other passive; the one is an occupation for a group, or at least two people; the other an occupation for the self.

It is true, of course, that people go to stadiums and theaters *en masse;* but they are, really, members of what David Reisman called "the lonely crowd." They sit in the audience, private and alone, interacting with the people around them only superficially and intermittently; and in many cases, as in a half-empty theater, interacting with nobody at all.

It is precisely this that makes the qualitative difference between playing the game and watching it: the spectator reacts but he does not interact.

Being entertained is essentially a solitary pursuit, an act of the self. It does not often involve us, or affect us, on a deep gut level, the way competing on a tennis court, playing poker in a basement den, jamming on the blues in a garage, telling ghost stories around a campfire do. These things engage us in a much deeper way, because they connect us to other people whom we can affect and who can affect us in significant ways—in winning and losing, in the rapid exchange of ideas, the sharing of an experience. The primary environment in which human beings live is other people. Playing tennis or jamming on the blues is life, because these are peopled activities. Watching someone else do these things is, at best, only an imitation of life. At worst, it is a way only of passing time in a state of disengagement.

And we are hardly surprised to discover that this basic shift in the way people amuse themselves came hand in hand with the new industrial city.

CHAPTER 9

New Thought, New Art

The changes in American society in the decades just before 1900—industrialization, the explosion of the city, the appearance of a huge new entertainment industry, the creation of an ever-growing professional sports system—were paralleled by sweeping shifts in American intellectual life. Artists, writers, philosophers, social and political thinkers were fervently, almost feverishly, pressing forward new ideas about how art should be done, society should be organized, life should be lived.

The new ideas were not random. Whether philosophy, or social science, the new ideas were running in the same direction: they were calling for spontaneity and open expression of feeling instead of Victorian self-control. They were demanding that the visceral rawness of life be exposed, a cry exactly contrary to the Victorian idea of modesty and concealment of the rough underside. The new art, the new literature, the new philosophy were determined to throttle Victorianism—and it would be done with glee.

Perhaps the earliest, and among the most important, manifestations of the new thought appeared in the ideas of the pragmatist philosophers, the best known of whom, in America at least, were William James at Harvard, and John Dewey at Columbia. (It should be pointed out that James drew heavily on the thought of the then unknown Charles Sanders Peirce.) To simplify drastically: for James it was experience, not over-arching, immutable laws, that provided a guide to "the truth." According to the philosopher A. J. Ayer, ". . . we may infer that he [James] meant to analyze one's conception of an object . . . in terms of the difference to one's sense-experience which the object's existence or non-existence would be expected to make."[1] For James, furthermore, the mind was not a fixed substance, but, says Ayer, "constructed out of

105

items of experience."[2] And when it comes to the moral questions, again James refers to experience: how does the belief work for us?[3]

To many of his critics it appeared that James was insisting that you could believe whatever it was useful for you to believe. Actually, as Ayer was quick to point out, James's views were not that simple.[4] But this was how, to a considerable extent, his ideas were perceived as they filtered out to the public.

James published much of his major work in the first years of the 20th century, and in those same years John Dewey was also having a dramatic influence on the way people, at least educated people, were coming to view the world. Dewey's impact was more direct, because he was deeply involved with the philosophy of education, and ultimately his ideas became the basis for a great deal of educational practice, the effects of which were felt beyond the middle of the 20th century. Basic to Dewey's philosophy was the idea that education is growth. The school ought to be centered on the child rather than subject matter, and the child ought to be allowed to explore for himself the world around him, rather than simply be told what he ought to believe. "The effect of this idea," says Hofstadter, "was to turn the mind away from the social to the personal function of education; it became not an assertion of the child's place in society, but rather of his interests as against those of society."[5] He adds that Dewey believed that education had a social as well as an individual function: the child was to be "saturated with the spirit of service." But it was the individual function that attracted his followers, and they pushed it at the expense of the social function.[6]

By 1900 Dewey's ideas were being adopted by educators everywhere in the United States. One said, "We have quit trying to fit the boy to a system. We are now trying to adjust a system to the boy."[7] Out of the ideas came the whole notion of "life adjustment" as the goal of education, which came to dominate the American school through much of the 20th century. The function of the schools was no longer to impart a body of knowledge but to help individuals be happy and productive.[8]

James and Dewey, both psychologists as well as philosophers, were attempting to build their ideas of what human beings should, or could, be, on some realistic understanding of what they were. But it was three other psychologists whose ideas would have the greatest impact on the public mind, and provide powerful weapons for the intellectual attack on Victorianism.

The most famous of these is, of course, Sigmund Freud, who first became known in the United States as a result of his lectures at Clark University in 1909. Freud's thought is complex and contradictory, and he would have been among the last to deny the existence of "instincts." But basic to his psychology was the idea that upbringing often caused people to "repress their emotions," and that a repressed emotion was likely to surface as anxiety, depression, or actual physical illness.[9] It was this aspect of Freud's thought that people, especially the young intellectuals of the early decades of the century, took up. One ought not to repress one's emotions; one ought to express them—and this was especially true of the sexual drive, which was basic to much of Freud's theory. Sigmund Freud was by no means saying that people ought to go around doing what they felt like doing; he was, at bottom, too much of a Victorian pessimist to believe that, and the whole point of *Civilization and Its Discontents* was precisely that society does not permit us to do everything we want, and thus places limits on our happiness.[10] But the notion that happiness was out there to be grasped if only people freed themselves from the repressive influences visited upon them in childhood was a far more welcome idea.

A second psychologist who contributed to the belief that people were born as blank slates was Ivan Petrovich Pavlov, a Russian who did important work on the digestive system in the late 19th century. Pavlov showed that he could condition a dog to salivate whenever a signal was given. This seemed at the time a bizarre idea; Pavlov's work was "good copy," and was written up extensively in the popular press. Like the work of Freud, it seemed to suggest that organisms would be shaped in almost mechanical ways by "conditioning."[11]

Both Freud and Pavlov remain well-known names. Yet a third psychologist, forgotten today by everybody but specialists, was more influential at the time. He was John B. Watson, considered to be the founder of the behaviorist school of psychology, which remains immensely important not just with psychologists but with the general public.[12] Watson believed that there was very little, if anything, like an inborn human nature. People were formed in their early years by responding to stimuli, which tipped them in one direction or another, forming their characters. Watson's ideas also made good copy, and thereafter the American public was deluged with behavioristic ideas on life, and especially on child-rearing.

Implicit in the ideas of James, Dewey, Freud, Pavlov, Watson, and other thinkers of the time was the idea that human beings were far more malleable than the Victorians had seen them to be. The Victorians did, of course, believe that human beings had to be shaped as children, and even as adults; people had to learn to control their baser natures. They must, as Daniel Walker Howe puts it, achieve "mastery over the 'bad passions' within oneself."[13] John R. Reed, another student of the era says, "Early Victorian sentiments had assumed a human will driven by passions that must be restrained to harmonize with the divine plan."[14] People could learn to control their passions; but it was assumed that these baser passions were a fixed part of human nature.

Dewey, James, and their followers made no such assumption. They held a much sunnier view—that children could be raised to be good citizens as easily as they could bad ones. Indeed, there was a sense among these thinkers that a child, if allowed to bloom naturally, would certainly be good.

This shift away from the belief in a basic human nature to a view of the human as the famous blank slate on which anything could be written—"the malleable, undetermined nature of man," as Henry F. May puts it [15]—is one of the most important—and perhaps *the* most important—changes of mind in modern times. It has been the basic belief on which many, if not most, of the institutions of modern America are built. Behind our philosophies of education, crime and punishment, poverty, and much else lies the idea that people's behavior can be shaped by the conditions in which they find themselves. Slums breed crime; the answer is to pull down the slums and replace them with decent public housing. Children become delinquent because of poor parenting; the answer is to supply children with professional substitute parents through Head Start, day care, earlier schooling. Teenagers become delinquent because of "bad home conditions"; the answer is to place them in shelters and homes where conditions are good. The idea, that people can be shaped, not only as they grow, but later in life, by the environments they live in, is basic to the beliefs of most Americans.

We must understand that few of the psycho-philosophers we have been discussing believed that the human being was malleable on every point. Freud in particular saw underlying instincts in people, and most of the others would have said there were limits to the extent to which humans could be shaped. Furthermore, their work was frequently inconsistent, and filled with counter-currents, complexities, and obscurities. Inevitably, the public got a vastly over-simplified picture of this thought—or rather, the public did what it usually does, which was to take from the complex works of these thinkers the few ideas it suited them to take. They saw the new thought as saying that they had been enchained by their Victorian upbringings, which had twisted their natures with neuroses and complexes. They must now break these internal chains, in order to express themselves fully, to enjoy their lives.

But however much the public oversimplified the ideas of these thinkers, they were quite right in seeing that the new thought had shifted the focus from society to self. These thinkers were not particularly interested in reforming society, although none of them thought that you could do without that abstraction; they were suggesting that human beings could be *re-formed*, through psychoanalysis, conditioning, or whatever, in order to make them happier and more productive. And this in turn was bound to have beneficial effects on the society.

The Victorians, too, had exerted great efforts to shape people; the difference between the two systems of thought lay in the new idea that the best way to make people "better" was to allow them to be more fully alive, to "express" their basic feelings. There is, of course, a contradiction here: if people can be shaped, it is just as reasonable to change

their basic feelings as to allow them to express them. Or to put it another way, if people can be anything, how do we decide what they *should*
be? That is to say, a person who is amorphous at birth, and subject to
revision throughout life, does not truly have a *self*: what shape is water?
But in the general enthusiasm for what appeared to be scientific support for the idea that you could—indeed should—do whatever you felt
like doing, this contradiction escaped notice. The self should be expressed, the self could be shaped—either way, the self became the focus of attention.

The ideas of Dewey, Freud, and the others were frequently recon-
dite, and it took a while for them to filter into public consciousness.
But they were just the sort of thing that magazine editors liked to give
their readers, and through the beginning decades of the twentieth century they gained currency, until by 1920, and perhaps earlier, it was
widely understood, in the middle class in any case, that it was dangerous to "repress your feelings"; that children ought not to have their
budding personalities smothered; that everybody ought to behave as
"naturally" as possible. (Actually, the view of John B. Watson, so far as
child-rearing went, carried him to an opposite corner: as the child was
to be shaped through conditioning, his desires should not be given into,
as that would only reinforce them; let him cry in his crib and eventually
he will stop wanting whatever it is.) The result, in the end, was that
"self-fulfillment" came to be seen not simply as a worthwhile goal but a
birthright.

Precisely at the same moment that Dewey, Watson, and the others
were advancing the new psycho-philosophic ideas, there were churning
through painting, music, literature, and the other arts very similar currents. In this instance, many of the new tendencies were being taken
from a group of somewhat older European artists, like Ibsen, Shaw,
and Wilde, who began in the main attracting the attention of intellectuals and the literate public in the 1890s, both in Europe and the United
States. Ibsen in particular was consciously dealing with the conflict between the individual and society, with his sympathies clearly with the
individual; and both Shaw and Wilde were writing about the hypocrisy
which appeared at that late date to be endemic to Victorianism. Somewhat earlier other Europeans, like Tolstoy in *Anna Karenina*, Balzac in
Nana, Flaubert in *Madame Bovary*, and Hardy in *Tess of the D'Urbervilles*
had attempted to write with greater specificity about sexual relations
than the Anglo-American Victorians generally allowed, and in the United
States by the 1890s they were being followed by many educated people.

In France the Impressionist and Post-Impressionist painters were
creating a new, and so they thought, scientific way of looking at the
world as consisting, for the painter's purpose in any case, of light. They
were also replacing the classical subject matter of the Victorian salon—
all those Venuses and Bacchic orgies—with the brothels, bars, ballet

schools, café-concerts of the modern city aborning. The poetry of Verlaine and Rimbaud, with its themes of depravity, was causing a stir in Bohemian circles. The idea now was not that art must uplift; it was that art must reflect real life. And the Victorian life many of these artists saw was filled with contradictions.

American writers, painters, sculptors, poets, were, by the 1890s, well aware of these European currents. Painters tried to find ways to spend a year or two in Paris, where they were frequently caught up in the Impressionistic mode. People were reading Balzac, Ibsen, Shaw. And inevitably America began to develop its own version of modern art. Theodore Dreiser, a major figure among the American realist novelists, published *Sister Carrie,* on the theme of the innocent girl coming to the city, in 1900, thus confronting what was seen as a major social problem of the day. Stephen Crane's *Maggie: A Girl of the Streets,* which appeared in 1893, also depicted a basic social condition of the period. Neither of these books was a success—indeed *Sister Carrie* was attacked as pornographic and was withdrawn.[16] But Crane had a great success with *The Red Badge of Courage* in 1895, and again with *The Open Boat and Other Tales* in 1898. By 1911, when Dreiser published *Jennie Gerhardt* times had changed sufficiently that, although the book was attacked as immoral, it survived and gave Dreiser his first reputation.

One of Dreiser's greatest supporters and his close friend was Henry L. Mencken, who eventually made an enormous reputation for himself as a gadfly on society.[17] Mencken, although he had a shrewd eye for what he could get away with, was virtually in the business of banging away at Victorianism. He was open about his love for beer, a cultural residue of his German ancestry, and both he and Dreiser lived at times with women they were not married to. "Mencken and Dreiser were very conscious of working out a philosophically tougher version of realism than had been fashioned by earlier pioneers like Howells and Hamlin Garland,"[18] says one author.

In Chicago, in 1912, an aging spinster named Harriet Monroe fell under the spell of Ezra Pound, who enchanted a lot of people in the literary world of the time. Pound was already beginning to have a profound influence on the new poetry of the 20th century through his encouragement—and editing—of many of the major figures to come out of the first fresh blast of modern literature, among them T.S. Eliot, William Carlos Williams, William Butler Yeats, and many others. Monroe began publishing Yeats, Pound, H. D. (Hilda Doolittle), Richard Aldington, Vachel Lindsay, and her fellow Chicagoan Carl Sandburg.[19]

It was, in fact, a great moment for English literature: Eliot wrote "The Love Song of J. Alfred Prufrock" in 1911; in 1912 *Poetry* and the British equivalent *Georgian Poetry 1,* were founded; in 1913 Pound wrote to an impecunious teacher in Trieste named James Joyce asking to see some stuff, with enormous consequences for the novel.[20] In Paris another expatriate, Gertrude Stein, was writing her experimental sto-ries—*Three Lives,* perhaps her best book, was published in 1909. An important part of the thinking of these people was a belief in the "doc-trine of spontaneity."[21] Stein, in fact, early in her career, experimented with automatic writing in which the hand was supposedly set free to run its own course.[22]

Realism in literature was paralleled by a movement in art, today known as the Ashcan School, which was also taking the life of the modern city as its subject matter.[23] Its central figure was Robert Henri, and it in-cluded as well William Glackens, John Sloan, George Luks and such of their pupils as George Bellows, Rockwell Kent, Edward Hopper, and Stuart Davis, all to become important figures in American art of the 20th century.

Many of these people were back and forth to Paris—George Luks lived abroad for a decade—and they were familiar with the work of the Impressionists. The found in it not only a new way of experiencing light, but, as Robert L. Herbert has pointed out in his study of the Impressionists, a new subject matter, the modern city.[24] These painters were of course enamored of landscape, and especially of water, which offered tantalizing problems of light to deal with—Monet's famous water lilies are the obvious case in point—but they poured enormous energy into painting modern life—in particular Paris and its suburbs. What they saw there is instructive, for again and again they present us with people in isolation, as for example the sad-eyed barmaid in Manet's famous "A Bar at the Folies-Bergère," who looks pensively away from the customer who inexplicably does not actually confront her, or in Degas's "The Place de la Concorde," with several people, among them family members, facing in different directions. Indeed, Degas could paint dancing classes crowded with people, virtually none of whom ap-pear to be interacting with anyone else. Degas, with his subjects so often crushed against the edge of the picture, or posed in awkward postures, seems to be offering us metaphors for the disjunction of modern city life.

Ashcan painters were not influenced by the Impressionists alone. They, too, were reacting to new currents in American thought. The group originally coalesced in Philadelphia, where several of its members were working as quick sketch artists for the newspapers, among them Glack-ens, Luks, Sloan, and Everett Shinn.[25] The experience was important to their work. In the 1890s, and especially in the later years of the decade, the lords of the new "yellow journalism," like Joseph Pulitzer and William Randolph Hearst, had discovered that pictures sold a lot of newspapers. However, newspaper printing technique had not yet evolved to the point where it could successfully reproduce photo-graphs. Therefore, they hired artists who would rush to the site of a burning building or a train wreck, make quick visual notes, race back to the office and hastily make a line drawing of the event in time for the next edition of the paper.

By the turn of the century the halftone technique had been developed well enough to allow the printing of photographs in newspapers, and scores of quick-sketch artists were put out of work. But in the meantime a whole generation of painters had been trained in the speedy rendering of scenes from real life—not only train wrecks and hotel fires, but little vignettes of ordinary life used to illustrate stories, romance, and items of gossip the newspapers ran.

In time the group drifted one by one to New York, where there seemed to be better possibilities for work. Their basic philosophy was expressed by Robert Henri, who told his students, "Stop studying water pitchers and bananas and paint every day life."[26] He added, "Work at great speed. Have your energies alert, up and active. Finish as quickly as you can . . . Do it all in one sitting if you can; in one minute if you can."[27] Says Bennard B. Perlman in his study of the school, "Although Henri advocated that all life was acceptable subject matter, his philosophy resulted in the belief that dilapidated slums and the non-privileged working classes were nearer to reality than the lush and extravagant existence uptown."[28] The eccentric George Luks said, "Who taught Shakespeare technique? . . . Guts! Guts! Guts! Life! Life! That's my technique . . ."[29]

This line of thought ran through all their work, and is obviously very close to the way Dreiser, Crane, Frank Norris, Upton Sinclair and other writers of the period were thinking. The Ashcan painters, however, did not find immediate acceptance for their position. At the time American painting was dominated by a highly academic style using classical themes replete with cherubs, sileni, and extraordinary nudes all done with an exquisite attention to detail and high finish. It is worth saying something about this. A central feature of this academic painting, both in Europe and America, was the idealized nude woman. Bouguereau was the acknowledged master of the style. These nudes were all small breasted, perfectly proportioned young women, or even adolescent girls, often drawn frontally but in such a way that the pubic hair had somehow disappeared, ostensibly shadowed, but actually simply not included. These nudes were unearthly, like no women who ever lived, and as such they were acceptable—indeed doted on—by several generations of Victorians, especially of course the men, who would not in mixed company use words like trousers, leg, or pregnant. To the modern eye they are immensely prurient, because they are all pneumatic body, to use Aldous Huxley's exquisite term, and no personality. They are empty of character in precisely the same manner that the centerfold women of another age are empty of humanity. They were objects meant solely for masturbatory fantasies.

And herein lies the explanation for the extraordinary furor that was raised when Manet exhibited the Olympia in 1863. The picture is of a nude lounging on a rumpled couch, with a serving woman behind her holding a bunch of flowers, presumably sent by an admirer. The woman was presumed to be a courtesan or prostitute, but the picture does not make that explicit. She is wearing a tiny smile, and her face has a look of blank, staring insolence, as if she were saying, "Just who do you think you're looking at?" She has humanity, reality; some critics even complained that her hands were dirty. What made so many people of the time uneasy was that she did seem to be a genuinely sexual creature. That was the challenge: you did not masturbate to Olympia: you were forced to make love to her or leave. In this picture Manet has done what he so often did, which was to raise a moral question and leave it unanswered. The final effect was to show how fraudulent the nudes of Bouguereau and the others were. The Olympia ripped the skin from Victorian prurience. Robert Henri was particularly taken by the Olympia, as well as other of the Manet masterpieces, and they were undoubtedly in part responsible for the forming of the "real life" philosophy which undergirt the Ashcan paintings.

Given that the Ashcan painters had placed themselves at odds with the Victorian idea of art, it is not surprising that they were slow to find acceptance. By 1906 John Sloan, then thirty-five, had not yet sold a painting,[30] and the others likewise had studios stacked high with canvases. They were, however, after the turn of the century, being cautiously shown by the Academy, usually fairly high up on a crowded wall.

Finally, in February of 1908 they organized a show of their own at William Macbeth's small gallery at 450 Fifth Avenue. There was press attention, some of it favorable, some scandalized; big crowds; and a few sales. William Macbeth, dazed and overjoyed, lauded the show as "a remarkable success."[31] The show later traveled to seven cities, and within a year the eight artists who had shown in it had become known to cultivated people. Sales, however, were still slow, if improving. But the artists were encouraged, and a second show in 1910 drew such large crowds that the police had to be called out to avert panic in the streets outside the gallery. Another show was projected for 1913; this time Arthur Davis, a strange, quiet member of the group, was put in charge. Davis had come to conclude that exceedingly important work was being done in Europe by a new group of painters unknown in the United States, and he insisted that they be given solid representation.[32] The only venue Davis could find that would hold a show of this size was the new 69th Regiment Armory on Lexington Avenue. Davis hung the American and French paintings in separate areas. The French included the Post-Impressionists, the Fauves, the Cubists, as well as some earlier paintings to show the line of development leading up to the new work. Among them were works by Cézanne, Van Gogh, Edvard Munch, Redon, Duchamp, Lembruck, and Brancusi. It was instantly apparent to everyone who knew anything about art that the Europeans had completely outdistanced the Americans: the game was over before it had started. When Robert Henri first visited the show and saw the French work, he could not speak, but stood staring at Davis in silence.[33]

The Armory show of 1913 was the most important art exhibition ever put on in the United States, and was a critical event in opening to the American public modern intellectualism. The show was an immense success, a subject of endless newspaper and magazine stories, jokes, cartoons, serious critical articles and much noisy talk. It is hardly the case that American collectors began immediately buying Cézanne and Gauguin, nor did American housewives start putting on their kitchen walls Toulouse-Lautrec posters and Van Gogh sunflowers. That would come later. But the exhibit made modern art in America.

It also left the Ashcan painters relegated to seats in the back of the hall. We still value their work: John Sloan's street scenes, George Bellows' "Stag at Sharkey's," are well known to Americans. But it was the Cubists, the Fauves, Cézanne and Van Gogh, who came to dominate 20th-century art. However, the Ashcan painters opened the door through which Van Gogh and the others ran; they had fought the Academy at considerable personal sacrifice, and we owe them something for that.

It was clearly recognized at the time that the new art represented a triumph of the individual over society. The French Impressionists had earlier recognized this, and they "wished to avoid traditional academic conventions because they would link [their] art to a communal school rather than to a unique temperament."[34] The artist must be unfettered by convention, must be free to express his own ideas. In the United States after the Armory Show, Kenyon Cox, a critic hostile to the new painting, saw in it "an exaltation of the individual," according to Martin Green.[35] And he quotes the *New York Times* review of the show: "It should be borne in mind that this movement is surely a part of the general movement, discernible to all of the world, to disrupt, if not destroy, not only art but literature and society too . . . The Cubists and the Futurists are cousins to anarchists in politics."[36]

The Ashcan School was replicated in a school of photography that had similar aims. Its leading figures were Alfred Stieglitz and Edward Steichen, who "had been paralleling the fight of the New York Realists with their attempt to gain recognition for photography as an art . . . Their unposed pictures of immigrants and views of New York City street life cause opponents . . . to label them the Mop and Pail Brigade."[37] Stieglitz and Steichen had been showing a few of the new artists in a small gallery they had established in 291 Fifth Avenue, which became known simply as 291. Steichen, who lived in Paris for a period, was particularly responsible for calling to Stieglitz's attention people like Matisse, Picasso, Rodin, Braque, and others. Finally, in 1910 Stieglitz and some confreres were invited to arrange an exhibition of photography at a gallery in Buffalo. "The Buffalo exhibition marked a turning point," says one author; photography would no longer imitate painting, but would find its own way.[38]

There are striking resonances between what was happening in the visual arts and in formal music during the same period, as Eric Salzman has pointed out in his standard work *Twentieth-Century Music,* in particular the relationship of cubism to atonal music.[39] Among other things, the most important event in formal music of the period was the *succés de scandale* of Stravinsky's *Le Sacre du printemps* in 1913, the same year of the Armory Show in the United States. Salzman says, ". . . the moment when traditional tonality ceased to provide the fundamental expressive and organizational foundation of musical thought and was replaced by other modes of musical expression and organization . . . occurred in the years around 1900 . . ."[40] Three key works of the period—Stravinsky's *Le Sacre du printemps,* Debussy's ballet *Jeux,* and Schoenberg's *Pierrot Lunaire*—were completed in 1912.[41]

These events were taking place in Europe, but very quickly American composers began to adopt the new mode. By 1912 the American Charles T. Griffes was working in the new manner, showing influences of Debussy and Ravel in pieces like his *Symphony in Yellow* and *La Fuite de la lune,* according to H. Wiley Hitchcock.[42]

But the key figure in American music of the time, today widely considered the greatest of all American composers, owed little or nothing to the European avant-garde. Charles Ives was an original, a man who invented his own aesthetic, as Virgil Thomson later said.[43] "Ives worked far from the European centers and, in many cases, years ahead of his European contemporaries . . . He anticipated . . . just about every important development of the last sixty years."[44] Basic to his work was the use of two or more contrasting themes or even whole compositions running along together to create a highly dissonant, thickly textured work. He was intensely American, and constantly threw patches of marches, hymns, folk songs, and the like into his pieces. His work was tonal, bitonal, atonal by turns, and sometimes all of these at once. Disorder was intrinsic to his method—or rather, it was a natural and welcomed result of his lack of method.

Ives was born in 1874, and by the 1890s was composing hymns and chorales of various kinds for churches in which he played the organ. However, his major work came in the years from the early 1900s to the early 1920s, when he began to wind down for reasons of health. Much of his most renowned work was done in the years around the First World War: *General William Booth Enters Into Heaven* in 1914, *A Symphony: New England Holidays,* completed in 1913, and the *Fourth Symphony,* written between 1909 and 1916.[45] However, there are important works from the late 1890s right through to the 1920s.

Ives accepted from the beginning that his work was difficult and would not support him, and he worked in the life insurance business in order to support himself. He was hardly known before 1920,[46] and the Fourth Symphony was not performed until 1965.[47] He was not part of the intellectual currents of the time, tending to look back to the Transcendentalists, rather than to the thinkers of his own time, like Dewey and James.[48] It is therefore all the more striking that he should come along just when he did, creating his oeuvre at the same time that the cubists in Paris, and the Ashcan painters in New York, were offering their novelties. In an essay on Emerson, Ives wrote, "His underlying plan of work seems based on the large unity of a series of particular aspects of a subject, rather than on the continuity of its expression. As thoughts surge to his mind, he fills the heavens with them, crowds them in, if necessary, but seldom arranges them along the ground first."[49]

This is a very cubist idea. What we are seeing in the new painting and the new music was not just a turning away from classical themes to vernacular subject matter, but the use of a looser, less tightly webbed set of relationships. There is less forward motion in the music, less of the old idea of theme and variation; less precision and balance in the new painting than had been true of the highly formal work of the academicians. It was somewhat more arbitrary, somewhat more random, and indeed it reached the point in the aleatory music of John Cage and others, where portions of the composition were arranged according to the throwing of dice. Salzman says, "In Debussy's earlier work, the simple and classical patterns of contrast and return still govern the large forms; later even these vanish, to be replaced by ongoing associative forms which depart from one point and, without the necessity of substantive recapitulation, arrive at another. In a sense, this non-narrative, non-cyclical form—achieved by Debussy in a work like the ballet *Jeux* of 1912—represents the larger intellectual tendency of the pre-World War I revolutions which we call 'atonal.' "[50]

Not surprisingly, there were appearing in dance many of the same tendencies that were manifest in painting, photography, and music. In the 19th century there had come into existence in America a few schools which offered rough and very simple training in ballet, the purpose of which was, in the main, to produce small corps de ballet to accompany renowned dancers from Europe. These dancing girls were considered little better than prostitutes by many of the Victorians around them. For most people in the late 19th century, dance was what you saw in vaudeville and elsewhere—usually solo dances like the Highland fling, the clog, Irish jigs, and various derivatives developed, frequently, in the saloons of the vice districts.

Then, very late in the century, a group of women began to create, almost out of empty air, a new kind of solo dancing which owed something to vaudeville, but had pretensions to art. First among these was Loie Fuller, who was followed closely by Maude Allan, Isadora Duncan, and Ruth Dennis, who in time changed her name to Ruth St. Denis. Although Fuller inspired them all, they developed their methods independently, and by the early years of the 20th century they had become celebrated in Europe for their "expressive" dancing.[51]

These dances tended to be Greek or "Oriental" in tone, with the dancer garbed in a flowing white gown, fairly elaborate headdresses, ankle bells, and the like. Frequently they were meant to produce the effect of Greek statuary in motion.

But whatever the form, the basic idea was for the dancer to unleash her imagination and express herself however she was moved. According to Elizabeth Kendall, in her study of these dancers, "Their minds raged against the confinement of a corps de ballet, yet their bodies had been taught harmony. They were ballet girls possessed of those 'unballetic' qualities of the new American woman—her 'great fund of life,' her self absorption, her fearlessness, her unbounded imagination, and her often monstrous daring. It was the combination of the ballet girl's body with the American untamed spirit that produced our first solo dancers and our first native art form."[52] It was all there: the insistence on the natural, the spontaneous expression of feeling, the desire to make everything anew.

All of these dancers had their first successes in Europe, but by the season of 1909–10, St. Denis, Duncan, Allan, and Fuller had shows in New York. There quickly developed a vogue for this new kind of expressive dancing. Society women took it up and by 1913 the public had become sufficiently aware of it to make parodies of it possible in vaudeville.

Although some of these solo dances were supposed to tell a story, in the main they were meant to set a scene or a mood. In 1913 St. Denis told *Vanity Fair,* "My new dances will be fantastic little things with tone and color and intimacy in which rhythm and the poetry of motion will dominate."[53] It was all of a piece with Stravinsky, Gertrude Stein, *Les Fauves.*

One pertinent aspect of the new painting, the new music, the new dance was the extent to which it moved away from a time scheme involving past and future, to one that was much more an on-going present. Art music through the 19th-century Romantics had a temporal structure which, at least in the larger forms like opera and the symphony, was built on contrasting themes, recapitulation, climax, and contrasts of intensity, dissonance, and the like. The goal of much of this was to create expectation which could then be satisfied—or not, as the composer chose. Such music could not be enjoyed without the memory of what had gone before, and the expectation of something novel, but appropriate, to come.

Similarly, in much painting before the Impressionists the viewer was presented with a moment in a story carefully chosen to suggest both what had gone before and what was going to come. For most ordinary viewers the painting made sense only as part of the story, and the implication of past and future in it were part of the experience of looking at it. Anecdote was a major trait of Victorian painting.

Again, in the well-made Victorian novel everything was tied together: scenes were prepared for by previous scenes, characters were involved with other characters, and everything contributed to the flow of the story toward its conclusion.

But the new art was different. The music of the 20th century, as Salzman said, was to a larger extent "associative"—what arrived was not necessarily foreshadowed. There was less need to know what had gone before, less expectation of what was to come. Similarly, the Cubists seemed to be painting an endless present. The classic example is Duchamp's "Nude Descending a Staircase," the most widely discussed picture in the 1913 Armory Show. The painting gives us the figure of a woman repeated in a number of postures as she walks down a flight of stairs: the whole story has been shoved into the present. And in the novels of Joyce, Stein, and others, people, places, ideas frequently bobbed to the surface and then disappeared without leaving any implication. And necessarily, a dance intended to set a mood is inevitably an endless present.

This is not to say that relationship disappeared from art altogether; a totally random construction is not comprehensible. Many of the people in Joyce's *Ulysses*, for example, are significant to each other; Ives usually worked from the standard diatonic, tonal system, and practically the whole point of Cézanne's later work was to show how objects, color, and tones related. But this modern art was far more random than anything that had gone before.

What we are seeing in this new art, then, is very similar to what we saw happening to the people in the new industrial city. The question is, are these phenomena related, and if so, how? It is a crucially important question, and I will address it shortly. For the moment let us turn our attention to another critical movement of the early years of the 20th century in America that has had important ramifications elsewhere in the world—feminism. There have been feminists through history, but as an organized force, feminism came into its own in the Victorian Age, and particularly in the second half of the 19th century. Although it would seem to be antithetical to Victorianism, it was in fact a corollary to it, indeed almost an inevitable consequence of it.

As we have seen, one of the important shifts in attitude that took place with the rise of Victorianism in the early part of the century was a change in the way women were viewed. The 18th century saw them as devious, sexually voracious beasts; the Victorians, on the other hand, elevated them to the status of angels, and demanded of them a rectitude that it did not ask of men. Once this idea was ground into American consciousness, it became obvious that, if women were morally and spiritually superior to males, they ought to play a larger role in setting the tone for society than they had been allowed. It came to be believed, among other things, that if women were given the vote they would inevitably vote for reforms; the female vote would surely pull down the corrupt big city political machines that had arisen in the late 19th century on the votes of the workers.

The so-called woman movement that arose in the latter decades of the 19th century was purifying at heart. Women took the lead in the drive against the saloon, in the movement to save prostitutes from their fate, in the drive to close the vice districts. All these issues became intertwined with more directly feminist objects, such as suffrage, the right to hold political office, the right to enter business and the professions. Nancy Cott, in her highly regarded study of 20th-century feminism, says, "As much as nineteenth-century participants and observers oversimplified by speaking of *the* woman movement, however, that language spoke a substantial truth. The rubric acknowledged that discussion and demands and actions raised by women constituted an integral spectrum crosscutting other political and intellectual views, even when indebted to them. The individuals and intents involved, although analytically distinguishable, also intertwined and overlapped. They shared and forwarded the perception that the gender hierarchy of male dominance and female submission was not natural but arbitrary."[54]

This should not obscure the fact that the movement was filled with tensions and conflict. Various feminist groups had their own priorities—the vote, abolition of slavery, class struggle, the anti-liquor cause, sexual freedom, and more. Furthermore, some groups and individuals wanted only relatively modest reforms—child labor laws, curbs on prostitution, better prenatal care for working women. Others wanted to overturn the whole system and replace it with one more egalitarian and freer. These more radical groups carried their ideas to the point where, instead of being an extension of the Victorian ideal, they were anti-Victorian. But these factions joined at so many points, and were basically traveling in the same direction, that they reinforced each other in the effects that they had on the social system.

Basic to the feminist movement was an attack on the double standard. If anything is clear about Victorianism it is that it espoused an ideal that few people could live up to. The ideal existed for both men and women, but, as Victorianism settled in, it became understood that males would have a harder time obeying its strictures than females would. Especially in the post-Civil War period, when the growing industrial city was twisting the old culture out of shape, did the burden of carrying the Victorian ideal fall increasingly on female shoulders. Males were forgiven for sowing their wild oats, were understood to need masculine domains where they could repair to drink, smoke, and tell smutty jokes.

For a woman to do any of these things was to label herself as coarse and perhaps a prostitute. Males could vote, women could not; males could bull their way into public office, positions of power in industry, gain celebrity in sports, write for newspapers, become doctors, philosophers, and playwrights. Women could do none of these things.

What the 19th-century woman movement wanted was, at bottom, equality with males. Inevitably, it was to a considerable extent focused on the activities of men. In some cases the people in the woman movement wanted males to stop doing what they were doing; in other cases they wanted woman to be able to do them, too.

The new feminist movement that emerged in the early years of the 20th century was different. Cott says, "Only a rare quirk prior to 1910, usage of *feminism* became frequent by 1913, and almost unremarkable a few years later."[55] This was developing at precisely the moment that the ideas of Dewey, Freud, Watson, and others were having their first impact on American intellectuals, and it is hardly surprising that the new feminism was in some aspects different from the earlier woman movement. It was not reformist in the old sense of getting the men out of the brothels and the saloons. Instead its primary concern was women. We can get some idea of what it meant to women by looking at the words of some young feminists quoted by Cott:

> Here [the feminist] comes, running, out of prison and off the pedestal; chains off, halo off, just alive woman . . .

And she would have:

> a better time than any woman in the world ever before.[56]

And:

> I am a poet, a wine-bibber, a radical; a non-churchgoer. I will no longer sing in the church choir or lead prayer meetings with a testimonial.[57]

Feminists were described by Randolph Bourne, a radical journalist of the time:

> They shock you constantly . . . They have an amazing combination of wisdom and youthfulness, of humor and ability, and innocence and self-reliance . . . They are of course all self-supporting and independent, and they enjoy the adventure of life.[58]

The new feminism of the new century was about the self and freedom more than it was about the causes that had driven so much of the

19th-century woman movement. Twentieth-century feminists continued to have social concerns: they fought for female suffrage, and a
good many of them were socialists, trade-unionists, and the like. But
their basic demand was to be free from the sexual constraints, religious
obligations, masculine dominance, limits to emotional expressiveness
that their mothers—so they believed—had been, if not destroyed, at
least damaged by. It came on very rapidly in the years around 1910,
and in 1913 it burst into clear view, a subject being written about for
and against by magazines and newspapers across the United States. By
1913 feminism as an idea could not be dismissed: you might like it and
you might not—the majority of men, probably, did not like it—but it
had to be dealt with. For a young woman it posed a lot of questions
about sex, vocation, education, liquor, smoking, marriage, men: almost
everything she had been taught as a little girl was being questioned by
the feminists. She must decide—or try to decide in any case—where
she stood.

Related to feminism, if not exactly a part of it, was a dramatic change
in dress, especially in women's dress. According to Valerie Steele, in
her recent study of fashion, "It is clear that the 'modern' style of feminine dress appeared in the years between 1907 and 1913 . . ."[59] Steele
credits the highly influential clothing designer Paul Poiret with responsibility for creating the new look, although others undoubtedly played
roles.[60] *Vogue*'s Paris correspondent wrote, "The fashionable figure is
growing straighter and straighter, less bust, less hips, more waist, and
a wonderfully long, slender suppleness about the limbs . . . The leg
has suddenly become fashionable."[61] And inevitably those involved were
using the same language the Ashcan painters, the feminists, the socialists, were using. Poiret said, "It was . . . in the name of Liberty that I
proclaimed the fall of the corset and the adoption of the brassière which,
since then, has won the day. Yes, I freed the bust."[62]

Everywhere the cry was for the emancipation of the self. Yet there
were running through American society of the period at least two currents which seem, at first glance, to be antithetical to the exaltation of
the self that was central to the new thought and the new art: socialism,
unionism, and allied movements; and a much smaller anti-modern
movement which rejected mass production, mass consumption, and the
industrial city in general in favor of a more rustic ideal.

Socialism, a European idea which gained force especially after the
1848 upheavals, had little meaning for 19th-century Americans, who
until late in the century were still a rural people with immense Western
lands to expand into, and who were, furthermore, preoccupied with
other issues. American socialism, like so much else, grew in the soil of
the industrial city. Many of the immigrant workers had brought its tenets from their homelands, and they provided much of the leadership

as well as the rank and file for the movement.[63] By 1900 the candidate for the new Socialist party, Eugene Debs, polled 96,000 votes; in 1904 it was 402,000; and in 1912, 901,000.[64]

These figures were alarming to many Americans. Oliver Wendell Holmes said, "When socialism first began to be talked about, the comfortable classes of the community were a good deal frightened."[65] But to intellectuals of the period struggling to find a way to alleviate the ills of the society—the poverty, the political corruption endemic to the cities, the ruthlessness of the business leaders—socialism was a very attractive idea.

In the end the socialist threat to the comfortable classes petered out. The Socialist party got its highest percentage of the popular vote in 1912, and that was only about 6 percent.[66] But socialism was an important component in the thought of the intellectuals of the period, and it continues to have a vigorous life in academic circles to the present day.

Socialism was essentially communitarian, and as such ran counter to the new emphasis on the self. But it was also anti-Victorian: for one, it stood in opposition to the established political, religious, and ethical structures of Victorian society; for another, with its followers comprised mainly of immigrants, it did not accept the Victorian ethic of gentility and high morality. In any case, the intellectuals saw no contradiction between socialism and the new thought: as Martin Green has made clear in his study of the period, the same group of Greenwich Village intellectuals backed both the famous Armory show, and the equally famous—at the time—pageant put on by striking Paterson, New Jersey, silk workers.[67]

The second of these counter-currents, anti-modernism, was a sort of "back to the roots" or "back to nature" movement. Influenced by the Englishman William Morris, this philosophy espoused handicrafts, rural living, and a kind of salvation through good workmanship. Insofar as this movement was a reaction to the industrial city, it was certainly anti-modern; but in many respects it was of a piece with the new thought of the day. According to T.J. Jackson Lears, who has studied the antimodern reaction,

> By exalting "authentic" experience as an end in itself, antimodern impulses reinforced the shift from a Protestant ethos of salvation through self-denial to a therapeutic ideal of self-fulfillment in *this* world through exuberant health and intense experience.[68]

The anti-modern movement, thus, was at bottom part of the general movement toward the exaltation of the self.

In sum, during the first years of the 20th century a group of related ideas, theories, attitudes appearing in art, philosophy, literature, psychology, and elsewhere were being taken up by a young generation born, mainly but not entirely, in the 1890s. By 1910 or so many of these young intellectuals were finding each other in restaurants, saloons, cheap apartments in San Francisco, Chicago, New York, and other major cities. A group that had collected in New York's Greenwich Village by 1913 was in particular seen as at the leading edge of the avant-garde, but there were many similar, if smaller and less notorious, groups elsewhere. Henry F. May has described among these people what has been called the Liberation Movement, an informal philosophy that was vaguely socialistic or anarchistic and was demanding "a free and joyous life for all." It had a certain mystic element, was interested in spiritualism and "the Life Force."[69] Mabel Dodge, one of the dominant figures in Greenwich Village of the time, insisted that life must be lived spontaneously. "Let It happen, let It decide,"[70] she cried, a phrase that could have been shouted on a hippie commune fifty years later.

It is clear, then, that the artistic and intellectual movement of the period from the late 1890s through to the beginning of World War I was all of a piece. Painters, writers, photographers, poets, philosophers, sculptors were all taking their disciplines apart to see how they really worked. They were finding two great principles. The first was that art, philosophy, psychology, ought to be about real life. And it is clear from their work that they found more reality in the slum, the vice district, the saloon, than in a block of middle-class brownstones wherein the Victorian ideal was still fighting a rear-guard action. The underlying message was that Victorianism was not real. Sigmund Freud said it when he announced that beneath the Victorian surface was a storm of disreputable desires and drives. Theodore Dreiser said it when he wrote about the unacceptable sexual behavior of his characters. Robert Henri said it when he urged his students to draw life in the immigrant ghettos. John Dewey said it when he told teachers to let the child determine the nature of his own education. William James said it when he told his readers that truth sprang from feeling. The feminists said it when they shouted that they were charging out of prison with their chains and haloes off. Henry F. May concludes, "The [old] American credo . . . was not close enough to actuality, and it also failed to satisfy the darker, fiercer, more violent human emotions."[71]

The second great principle was freedom—in art, in life, in thought. At the time of one of the earliest of the independent Ashcan School shows, Robert Henri told a reporter, "Freedom to think and to show what you are thinking about, that is what this exhibition stands for. Freedom to study and experiment and to present the results of such essay, not in any way being retarded by the standards which are the fashion of the time."[72]

The people who were putting together *The Masses* took exactly the same tone. John Reed, the vastly admired journalist who was eventually buried in the Kremlin, said that the main point was "to do with the

Masses exactly as we please."[73] Art Young, an important contributor of drawings said it was to be "a magazine which we could gallop around in."[74] Joseph Freeman as a teenager saw the magazine as bringing "revelations of Maxist theory and of the possibility of a new way of living, 'frank and free.' "[75] Rebecca Zurier, in her study of the magazine said, "Freedom had become a form of challenge—to readers and staff alike."[76]

In the new world freedom was to be all: painters must be free, novelists and poets must be free, thinkers must be free, most especially to reexamine life and describe it in a new set of terms. And one effect of the new freedom was the abandonment in art, in ideas, in life, of fixed structures of relationships.

It is important for us to notice that the artists, who were mainly drawn from the middle class, were doing precisely what ordinary middle-class people, especially middle-class males, were doing—that is, turning to the culture of the proletariat for new modes of thinking, behaving, and perhaps even feeling. The file clerk making a visit to the local vice district was looking for the same thing that the painter sketching a slum street scene was after—a freedom and expressiveness that neither of them found in his own culture. The vision was rose-colored, romantic, of course: the girl working sixty hours a week in the sweatshop could hardly claim the enjoyment of much freedom; and the ill-educated sixteen-year-old prostitute from a broken home did not possess many means of expression. But they did at least appear to be living closer to the bone, to more openly enjoy drinking and dancing and sex and the rest of it, and the middle class was responding.

Underlying all of this yeasty, widespread, intellectual ferment—the new philosophies and psychologies, the new art, the new fashions, the new literature—was the idea that the needs of the individual should come first. It was an idea, certainly, that some of these people would have hedged with restrictions. Freud, Dewey, many of the socialists, and others would have admitted that unrestricted freedom was an unworkable premise on which to build a society. But a surprisingly large number of the intellectuals of the early 20th century did not make these exceptions: to many of them the individual must always be free of social restrictions. In practice, of course, few of them could actually live wholly spontaneous lives acting on the impulse of the moment. But many of them believed they ought to, and a lot of them tried.

The new ideas that were growing in intellectual and Bohemian enclaves inevitably began quickly to leak out into the consciousness of the populace. "Everywhere in the country in 1912 young people were getting restless . . . The restlessness usually arose from a contact between new ideas and traditional, even obsolete environment,"[77] May says. It was a tremendously alive and exciting period in art, ideas, politics, social theory, a time when it seemed that the new social sciences would solve the problems of the cities, a new art would bring an expressiveness that had been missing from American life for generations. It was a time when it seemed that America was being reborn and the world, as it entered the 20th century, would all be new. Floyd Dell, one of the leading writers of the Greenwich Village group, said later on, "Something was in the air. Something was happening, about to happen."[78]

CHAPTER 10

Hedonism Victorious

At this point it might be worthwhile pausing for a moment to look back over the road we have been traversing. We have seen, especially in the latter decades of the 19th century, a number of dramatic changes in American society. The first of these was the rise of the industrial city, built on technological innovation and a huge laboring mass flooding in from outside. This new city made possible, and perhaps inevitable, certain institutions and conditions which grew inside it. One of these was the new class system, which divided the populace into blue- and white-collar classes. Another was the institutionalized vice district. Yet another was the enormous entertainment/sports machine, which by the turn of the century had a firm grip on Americans of all classes: "laborers who raised their income above the subsistence level used much of the difference on leisure pursuits,"[1] says one report, and very shortly the middle class joined them at the movies and in the vaudeville theaters. We have seen, finally, how the condition of the city itself helped to weaken family and community bonds, leaving the mass of the people, of whatever class, less firmly tied to specific individuals, and somewhat more like interchangeable parts of a machine.

Paralleling the emergence of this new physical system of buildings, machines, the work place, and the leisure domain was the coming of a new mental system—a way of viewing life, art, education, and the mind itself. The new thought, which arrived in the main after the turn of the century, was various and ramified, but running through it like a scarlet thread was the idea that the self should be less constrained than had hitherto been the case with the Victorians, concerned with self-control as they were. The effect of this new attitude was once again centrifugal, for it inevitably vitiated the sense of responsibility to family and community people felt, or at any rate supposed to feel.

126

The final result of this massive upheaval in the way people thought about things and did things was to turn the city, after 1900, into a battleground between the Victorians and the new wave. On one side was the middle class, and especially the older members of the middle class, now three or four generations old and operating out of a well-established sub-culture with its mores, folkways, speech patterns, and habits of work and leisure. On the other side was the more fragmented working class, with the support of many artists and intellectuals who saw in working people a naturalness and liveliness they did not find among their own people. The middle class, in turn, was supported by the farmers and small-towners in rural America, the ones who played critical roles in pushing through Prohibition. But although the Victorian idea was entrenched in rural areas, the actual battle took place where the anti-Victorians had established their beachheads: in the vice districts, saloons, movie theaters, and dance halls of the big cities. Let us now see how that battle was fought, and how it was won.

Surprisingly, one of the most important affrays in the fight was over smoking. Smoking now is seen as a minor element in American life, which should, most people think, be curbed as a threat to public health. It is thus hard for us to believe that in the early 1900s it was considered a significant part of a revolution—particularly for women an issue worth struggling over. In Victorian times male smoking was, if not approved, at least condoned. However, it was understood that men would confine their smoking to masculine environments—saloons, private clubs, baseball grounds. Jack Larkin notes that "tobacco use marked off male from female territory with increasing sharpness," as the Victorian Age came on.[2]

But after the turn of the century women, especially the educated middle-class women who filled the feminist ranks, began insisting upon their right to smoke.[3] Smoking, it was generally believed, was a simple and relatively innocuous pleasure—surely less damaging to the constitution than liquor. Through the early years of the century more and more women began to smoke, until by the 1920s Caroline Ware, herself a feminist, could write of "the disappearance of smoking as an issue."[4] Inevitably, with women now smoking, middle-class men felt free to smoke in their own homes, and cigarette smoking, in Victorian times thought to be a nasty habit which lowered even the men who did it, became endemic in American life. Cigarette consumption soared from about 4.2 billion in the first decade of the century to 24.3 billion in the next decade.[5]

What was true of cigarette smoking was even more true of drinking. Once again, it was a change in behavior by women that triggered a more general change in the society. Middle-class Victorians anathematized alcohol in general, but it was certainly believed that, however sordid a drunken man appeared, a drunken woman virtually forfeited her right to be in the company of decent people. It was accepted that a man in the privacy of his club or bar might have a drink or two, provided he went about it in a seemly fashion. But women were not supposed to drink at all.

It is therefore not irrelevant that the feminist of the time quoted by Nancy Cott should claim the right, among other things, to be a "wine-bibber." Drink, says Carolyn Ware, "became not only an avenue of escape but a symbol of defiance."[6] Not surprisingly, then, through the first years of the 20th century there was a general rise in the level of drinking in America. Per capita consumption of absolute alcohol, which had been hovering around two gallons through the latter decades of the nineteenth century, was at 2.60 gallons in the 1906–10 period,[7] an increase of about 30 percent. As important as the increase was the fact that a long-term trend away from hard liquor to beer was continued. This meant that the 2.60 gallons of absolute alcohol was diluted through an even larger volume of liquid, suggesting that more people were sharing in the pool of alcoholic beverages than hitherto.

It is especially significant that this increase in American drinking came in the face of a tidal wave of prohibition sentiment. Through the first decades of the century state after state voted in laws meant to limit drinking: between 1907 and 1914 eleven states passed dry laws.[8] Anti-drink propaganda poured out of the presses of the Anti-Saloon League during the first two decades of the 20th century, bombarding the public with tales of horror from the saloons. Nonetheless, despite the steadily growing legal and moral constraints, alcohol consumption in the period was higher than it had been at any moment since the Victorian ethic had taken hold.

This increase in the amount of alcohol being consumed in the United States may have had several causes: the immigrants, with their tolerant views of liquor, still flooding into the country; and a greater prosperity than had existed in the depression years of the 1890s. But neither factor can wholly account for an increase of this magnitude, and it is generally believed that a considerable proportion of the increase was produced by a growing interest of the middle class in alcohol.[9] There is no good statistical evidence for this, but there are good reasons for believing it. For one, as we shall soon see, the effect of the prohibition laws, for various reasons, was to make it harder for working people than for the middle class to get hold of alcohol. It is therefore unlikely that working people were drinking more per capita; and in fact they were probably drinking considerably less, because high percentages of the new immigrants of the period were Italians and Jews, who did not have the traditions of drinking that the Germans and Irish who dominated immigration earlier did. As a consequence, the middle class must have contributed substantially to that 30 percent increase in per capita alcohol intake of the first decade of the century. Given that the middle class did not represent more than a third of the population at most, it can be estimated that drinking among white-collar people increased by at least 50 percent, and may have doubled.

This idea is supported by the biographies and reminiscences of the young bohemians and intellectuals, such as the painters of the Ashcan School, and the avant-garde gathering in Greenwich Village and elsewhere, drawn mainly from the middle class, who were making liquor a regular part of their lives.[10] Prohibition was "a middle-class, Protestant, a nativist activity,"[11] but certain elements in it, espcially the intellectuals and the young, were running counter to the main current. Once again the middle class was adopting a working-class custom.

At virtually the same moment a similar, although less ferocious, battle was being fought out over a wave of new social dances. The dances of the 18th century, and continuing into the 19th, were "set" dances—reels and what are known today as "square" dances, which were done in groups, with the males and females not necessarily paired off. However, in the latter part of the 19th century "couple" dances, like the waltz, the polka, and the schottische, were imported from Europe, mainly by the rich, who were still looking to the Old World for ideas. By 1886 wealthy and fashionable Chicagoans were studying these couple dances at Bournique's dancing school.[12] Couple dances, which brought males and females into close physical contact, suggested sex, and they were generally opposed by the middle class,[13] although here again there were some who were more tolerant of these newer dances, provided of course they were danced at home at parties of peers.

Then, with an astonishing swiftness, after 1907, there swept through the country a group of new dances, most of them quite similar, which have been given the generic term, "trots"—the Turkey Trot, the Fox Trot, the Grizzly Bear, the Texas Tommy, the Buzzard Lope, and others, many with animal names. Trots were, at first, fairly vigorous dances that employed a good deal of hip and leg movement, and could be performed with the couples in close bodily contact.

According to Jean and Marshall Stearns in their book *Jazz Dance*, trots derived from Africa by way of the black and tans of the vice districts, especially those of San Francisco's Barbary Coast.[14] The Stearns offer little evidence for this contention, however, and at least one other hypothesis is available. It is clear that vigorous couple dances were being done in Europe in the second half of the 19th century. Pierre-Auguste Renoir's "Dancing at the Moulin Galette," of 1876, and his pair "Dance in the Country" and "Dance in the City" show couples dancing in close embrace. The art historian Robert L. Herbert has also discovered a wonderful drawing of dancers at the "Bal Mabille" in which the movements of hands and feet are strikingly like those which could be seen at a club today.[15] Nor were these dances decorous. Herbert quotes a contemporary description of the famous Masked Balls at the Opera in

Paris: "Do not flatter yourself it is a stately affair . . . It is a mad whirl-pool, wherein all that is graceful is cast away and unlimited license of attitude takes possession of the field. It is a salmagundi of all ages, classes, and conditions; an apotheosis of embracings, of whirlings, of jumpings." And Herbert adds, "A woman who mingled in the crowds here, unless very closely guarded by a man, ran the danger of having her costume examined by many hands." These affairs were patronized by the haut monde, but included ballet dancers, who were drawn from the working class, courtesans, and the like.[16]

By the 1870s similar dances were being done in saloons in the United States. One New York City reformer saw dancers "glide over inclined plane, and the dance is swifter and swifter, wilder and wilder."[17]

Blacks appear to have had a somewhat different dance tradition. An examination of several hundred pictures of black life of the latter part of the 19th century showed no examples of couple dancing of this kind.[18] The few pictures of blacks dancing showed either the cakewalk or solo dances, mainly boys dancing for pennies. The black dance tradition, as we see it today, is of the solo artist, like Bill "Bojangles" Robinson and Johnny Hudgins of the early part of the century, or their lineal descendants, the jazz dancers Gregory Hines, Bunny Briggs, and others today. Trots, so far as we can tell from pictures, appear to be more closely related to the European couple dances than to black ones. Given the number of Europeans pouring into the United States in this period, it is hardly surprising that they had an influence on American social dancing, as they had on so much else. We remember that the casino garden, featuring dancing, was devised by European immigrants on European models. Nonetheless, it is reasonable to assume that there was at least a measure of black influence at work in the development of trots, along with influences from older native dances.

But if we do not know who invented the new dances, we can be certain that they came out of the vice districts, especially the Barbary Coast. Irene Castle, who, with her husband Vernon, was the most famous show dancer of her time, said, "We get our dances from the Barbary Coast. Of course they reach New York in a very primitive condition, and need to be considerably toned down before they can be used in the drawing room."[19] In 1922 the *New York Times* said, "Here the turkey trot, the bunny hug, and the rest of the 'gutter dances' originated . . . The Barbary Coast is now a memory, but the rest of the Pacific Coast is jazzing [dancing] away to the latest of Broadway melodies."[20] Herbert Asbury, in his study of the area, said that in the period after the famous fire of 1907, "The principal attraction was dancing. The whole Barbary Coast was dance crazy, and practically every dive of any pretentiousness was a combination dance-hall and concert saloon . . . By 1910, four years after the disaster, there were no fewer than three hundred saloons and dance-halls crowded into six blocks, centering, of course, in Pacific Street."[21]

Trots were generally composed of a few quite simple steps, and were fairly brisk. Some were quite athletic. The Texas Tommy was danced with the couple side by side, clasping each other around the waist. They took two long strides and then two or three little hops, stamping their feet between hops, according to one contemporary observer.[22] But pictures mainly show couples facing each other, as is usual with couple dances, in as close an embrace as they wished.

By 1907 these dances began to break out of the vice districts and into the mainstream. Within two or three years the new dances had become a rage, a craze, a national phenomenon. Reformers everywhere were decrying them as wanton and destructive of the morals of the young. But there was no halting the flood. Particularly important in making the new dances respectable were the Castles. He was an Englishman, she the daughter of a Westchester County doctor who had become stagestruck. With their backgrounds they could not be dismissed as coming out of the lower orders. Furthermore, they were married and danced with a grace and dignity that made it difficult for reformers to criticize them. Gilbert Seldes, chronicler of the "lively arts" for the period, said, "There was always something unimpassioned, cool, not cold, in her abandon; it was certainly the least sensual dancing in the world."[23]

"By the fall of 1913," Irene Castle later wrote, "America had gone absolutely dance-mad. The whole nation seemed to be divided into two equal forces, those who were for it, and those who were against it . . . We were clean-cut, we were married and when we danced there was nothing suggestive about it. We made dancing look like the fun it was, and so gradually we became a middle ground both sides could accept."[24]

But if the middle class was taming the "jazz dances," as they were initially termed, working people were doing them in the old manner. As the dance boom rose, dance hall openings took place with the rapidity of machine gun fire. According to Kathy Peiss, "Of all the amusements that bedazzled the single working woman, dancing proved to be her greatest passion." A 1910 survey indicated that nine out of ten girls in the eleven to fourteen age group could dance, and about a third of the boys. Working-class organizations regularly put on "rackets," which might attract as many as a thousand dancers, mostly young, but the main venues for dancing were the commercial dance halls which sold liquor, where young working men and women—indeed boys and girls— could meet.[25] Sexuality abounded. Peiss quotes one contemporary observer as saying, "Couples stand very close together, the girl with her hands around the man's neck, the man with both his arms around the girl or on her hips: their cheeks are pressed close together, their bodies touch each other."[26]

According to the report of a New York City Committee on Amusements and Vacation Resources for Working Girls, made in the winter of 1908–09, admission ranged from five to fifty cents, with the boys charged a little more than the girls. The majority of the dance halls sold liquor fairly indiscriminantly, and the places were packed with teenage girls hoping to meet males. Inevitably, those of the middle class who were fighting to maintain the old ethic tried to constrain what they saw as the worst aspects of the dance halls. "There were found certain features, as in the freedom of attendance and the lack of restrictions as to the making of acquaintances, in nearly all of these places, which made them objectionable even in the absence of liquor." The committee did not object to dancing as such, for it felt it was a "natural" interest of young people; it was the ambience it took place in that bothered them. The report said that the saloon dance hall "is there for the purpose of selling liquor and offers dancing as a bait." Many had "a hotel attachment." "On an average Saturday night fully 2,000 unescorted young girls enter the dance halls of the city, in many cases without the knowledge of their parents."[27] Officially the city found all of this deplorable, but, as had been the case with the vice districts, enforcement was another matter. The Children's Aid Society raided a dance hall on Sixty-sixth Street and Broadway in New York City, where Lincoln Center now stands, and found eleven girls there aged eleven to seventeen. Captain Thompson of the Sixty-eighth Street Station "seemed much surprised by the raid. He said that his men had made inspections of the hall and found it was conducted in an orderly manner."[28]

What the New York group uncovered, however, was relatively tame. The Philadelphia Vice Commission investigator witnessed a Beneficial Society Ball, put on for a crowd that included a lot of prostitutes and toughs. "There was positively no restriction to the rough dancing, and I saw the most lewd and wanton actions by many couples on the floor. The turkey-trot, wiggle, spot dancing, and every kind of licentious dancing was indulged in. I saw unmistakable signs of sexual passions during the dancing and one girl I noticed was so worked up that she had to leave the floor and go into the ladies parlor. There was no question from her actions what the matter was."[29] In events like these there would be girls as young as twelve drinking beer.

A variation on the dance-hall was the taxi- or jitney-dance hall, where the males were charged a fee for each dance, usually a nickel or a dime. The taxi-dance hall had a relatively brief existence as an important American institution, but it entered into American myth through songs like Rodgers and Hart's "Ten Cents a Dance" and many movies, usually presenting the romantic tale of the poor waif forced to dance for a living. It has been suggested that the taxi-dance hall descended from Barbary Coast dives, like the '49 Dance Hall, where girls were provided free for dancing but were supposed to hustle drinks.[30] The taxi-dance halls attracted unattached males, once again including a substantial number of teenagers, who bought rolls of tickets. At the conclusion of each dance the girl took a ticket. Dances lasted 40 to 60 seconds and were followed by a 30-second intermission.

Like the open dance halls, the taxi-dance halls were primarily haunts of working-class youth. Working two or three nights a week, the girls could make double what they might make in the sweatshops, and there was, besides, what seemed to them the glamorous atmosphere. For the males the primary draw was sex, although certainly the enjoyment of dancing entered in. The boys tried to date the girls, and the girls teased the boys sexually to keep them on the dance floor. In fact, many of the girls did go out with the boys after the hall was closed; some entered into semi-permanent relations with them, with the understanding that the male would help with the rent and provide a stream of cheap presents. Others were quite simply prostitutes.[31] Yet these places, like the vice districts, attracted a considerable number of young middle-class males. "One reformer pointed out that dance halls became so attractive to working-class women because they offered the opportunity of meeting college men sowing their 'wild oats.' "[32]

By the war years, dancing had become a significant activity at all levels of the society. Factory workers danced during their lunch hours, and so did society women in restaurants and "tango places," where there might be instructors to dance with. Restaurants quickly learned they had to provide music, so that people could dance between courses. The tea dance, or *thé dansant,* as the more elegant would put it, held at five or six in the afternoon, became popular among the better off.[33]

The vogue for public dancing was an important factor in the appearance of another institution which typified the era—the cabaret. The first cabaret as such in the United States was the Follies Bergère,[34] which was opened in 1911 in New York's Broadway district by Henry B. Harris and Jesse Lasky, who two years later would help to found the movie industry in Hollywood and go on to have a major career as a film producer.[35] The place was supposedly based on the Parisian *cafés-concerts* the Impressionists so often painted. It offered dinner, champagne and an elaborate show. It quickly failed, because, Lasky said later, it did not offer dancing.[36]

But others seized upon the idea and began to produce lavish restaurants that offered formal shows and the opportunity to dance. A space was made in the middle of the room to provide a dance floor and a small area in which performers might work, giving rise to the term "floor show" as opposed to a show performed on a raised stage set apart from the audience.

This was a critical point. The entertainers and dancing couples were now close to the audience, in some cases close enough to touch. Lewis A. Erenberg, in his study of New York cafes, says, "The dance floor, the absence of large proscenium arch stages, and the closeness of the audience seated at tables made the room a scene of expressive activity.

The entire restaurant became the setting for the action and spontaneity of the moment. In the theater, expressiveness was limited primarily to hired performers. In the cabaret, audiences and performers were on the same level, and thus expressiveness spread to the audience as well."[37]

Patrons and cabaret owners liked to think that the cabaret, as its name suggested, was very Frenchy, and therefore elite; but in fact the institution owed more to the saloons and dance halls of the vice districts and elsewhere than to Paris. The trots that were done in them and the ragtime and jazz music that were danced to had come out of the vice districts. The Original Dixieland Jass Band was brought into the famous New York lobster palace, Reisenweber's, directly from a low Chicago club: Nick LaRocca, the cornetist with the group, said that the group had to buy dress suits for the occasion. "We were coming out of a dive and going into one of the most fastidious and largest places in New York."[38] It was not only the bands, however, which were being brought out of the dives into the sunlight of the fancy cabarets and restaurants. Many of the biggest cabaret stars, like Al Jolson, Sophie Tucker, and Jimmy Durante,[39] had begun entertaining in saloons, rathskellers, and casino gardens for crowds dominated by first- and second-generation immigrants. The Frenchiness was only on the surface: the entertainment, the behavior, and accompanying attitudes displayed in the cabarets were essentially home grown.

The cabaret was developed specifically and consciously as a place of entertainment for the middle class, and the upper crust. They were elaborately decorated, expensive, and demanded a certain tone—good manners, formal dress, and in particular careful control of mingling to keep men and women from meeting who had not been properly introduced. This concept of how people were introduced, and to whom they were introduced, was critically important to the wealthy at the top of society, and not much less important to the middle class a step or so down the ladder. It was crucial that the people who were brought into one's social circle were not adventuring males, or equally adventuring females, who might excite the interest of the marriageable in the group. Indiscriminate mixing, especially of the sexes, was seen as characteristic of the lower orders, something that went on in saloons and cheap dancehalls; the cabarets could not have attracted the people they wished to had they permitted it. People were therefore seated in enclosed groups at tables as they came in; you stayed with your party in the cabaret, ate with them, danced with each other. Women were not allowed into the bars that many of the fancier cabarets provided for men; indeed, some even created special bars where women could go alone, or with their male companions.[40] The essential point was that in no sort of bar were single people of *both* sexes permitted.

The function of the cabaret, and its relative the lobster palace, offering dancing and a show was to provide a place for the higher levels of the social system to enjoy the kind of good times that working people sought in the saloons and dance halls. Middle-class males of the period, through their visits to the vice districts, knew about those good times, and it is very likely that a great many of their wives and daughters knew more about them, through gossip, than they pretended to know. It is not irrelevant that the years after 1911, when Harris and Lasky opened the first cabaret, were precisely the period when the purity reformers were closing down the vice districts, and the prohibitionists were eliminating the saloons in state after state. Paul G. Cressey, in his study of dance halls, says, "Perhaps more than a mere coincidence is the fact that the first steps in the evolution toward the taxi-dance hall pattern occurred after the nationwide movement to wipe out the segregated vice areas in our cities, and were contemporary with the local-option campaign against the saloon and the enactment of the national prohibition."[41]

Neither prostitution nor the saloon was gone: the middle-class male could find as much of either as he wanted. But the kind of expansive hedonism of the vice districts, combining as they did music, dancing, liquor, female nudity, exotic and more general entertainment, and plain old-fashioned sex, was disappearing. For the middle-class male, the cabaret filled a growing void. What was happening, in fact, was the cabaret, the elegant dance palace like the Castle House with its *thé dansant*, and the "lobster palaces"—fancy restaurants—were replicating on a more mannerly level the style of the saloons, honky-tonks, and rough dance halls of the vice districts.

The institution could not have flourished, however, had not middle- and upper-class women felt comfortable entering. As it had been with smoking and drinking, so it was with what would eventually be called "night-clubbing." Infused with the new spirit of freedom, middle-class women, especially young women, were demanding the right to experience things they had not been permitted under Victorianism. The cabaret, in a sense, brought it all together: champagne, cigarettes, the new somewhat sensual dances, slightly bawdy entertainment, hot music. But it was only possible if there was no cutting across social class, for, however exciting the new freedom was, it must not lower a woman's status. For better or worse, the status of a woman was determined by the status of the males she was attached to, as father, brother, husband, and son. A man could keep a dancing girl as a mistress without lowering himself; a woman could not do the same. Erenberg says that Julius Keller, proprietor of Maxim's, "saw clearly that the transmission of cultural forms from the joints to the cabarets required a show of legitimacy and class if well-to-do patrons were to accept them. Hoping to attract a respectable clientele to his new Thirty-eighth Street establishment when he moved from his Tenderloin dive, he decorated with an explicit French theme and barred the most objectionable of his former patrons."[42] Said Keller, "After all, the things which appeal to the various strata of society are basically pretty much the same; the differences consist mainly in the degree of embellishments."[43] The cabaret, then, was really just a refined version of the old concert saloon.

Of all the changes in behavior that occurred in the period we are talking about the most basic one—one all the others circled warily around—the shift in sexual behavior. Unfortunately, it is far harder to document this change than it is to explore the new freedoms in the areas of drink, dancing, and night-clubbing. Of the huge amount published on sex in the past fifty years or so, only an astonishingly small fraction is of much use to anyone trying to get at behavior patterns.[44] For earlier periods the situation is even worse. There were a few early surveys of behavior, among them ones by Katherine Davis and G. Hamilton,[45] which studied small populations of women, some of whom had been raised in the Victorian period. The problem with all these earlier, and many more recent sex studies, is that they used samples heavily biased towards the middle class, usually self-selected. The consequence is that the best source for information on American sexual behavior prior to recent times is the famous Kinsey Reports, which appeared in 1948 and 1953. The Kinsey Reports, which are actually books entitled *Sexual Behavior in the Human Male,* and *Sexual Behavior in the Human Female,* have two great advantages: (1) they surveyed a large population of some 12,000 people, drawn from almost every section of the social system; (2) Kinsey in the main avoided the problem of self-selection. In finding respondents, he would ask a group—a P.T.A., a service club, and the like—to volunteer as a group—that is to say, everybody in the organization would be asked to contribute his or her sexual history.[46] Kinsey admitted that the study was tilted towards the middle class, and that insufficient numbers of blacks, laborers, and other groups had been covered.[47] Taken as a whole, the survey does not meet modern standards for demographic sampling. But if the population it surveyed was not completely representative of the society as a whole, it nonetheless was a fair approximation, with at least some members of almost all segments of the nation included. The Kinsey Reports, which gave us our first real look at a central human activity, are among the most important documents produced by the social sciences in this century.

For our purposes, the fact that they were done some two generations ago proves helpful. Kinsey was taking sexual histories as early as 1938,[48] which means that the bulk of his respondents were born between the Civil War and World War I. Some of them, thus, were Victorians, and it is from these reports that we get our best information on Victorian sexual behavior, and the shifts that came in the early part of this century. Unfortunately, Kinsey and his colleagues did not break all of the results down by birth cohorts; the results for the males were divided only into an older generation and a younger one; and although later generations of females are divided by decade of birth, all females born before 1900 are lumped, presumably because there were insufficient numbers of them to produce statistically significant figures. Nonetheless, given these limitations, an examination of the Kinsey Reports can give us a fairly good idea of how sexual behavior changed in the period we are looking at—the best idea, in any case, that we are ever likely to have.

What is most striking about the patterns for males is how little they changed between the two generations. We must be a little cautious about this. Kinsey divided his sample of males into those thirty-three or older when they were interviewed, and those younger. The median age of the older group was about forty-three years, which means that virtually all of them were born before 1913, and about half of them were born before 1900. "The older group represents the generation that was in its youth and therefore sexually most active from 1910 to 1925. These are the individuals who fought World War I and were responsible for the reputation of the 'roaring twenties.' "[49] (The point of this exercise was to check to see if the Jazz Age, only gone for a decade at the time Kinsey began his interviewing, had actually produced dramatic changes in behavior, as was generally believed.) To put it very roughly, then, we are dealing here with a younger generation of men who were the fathers of people now in their thirties and forties, and an older generation who were their grandfathers.

On most of Kinsey's indices there is little or no change in sexual behavior between these generations of males. "In general," the report says, "the sexual patterns of the younger generation are so nearly identical with the sexual patterns of the older generation in regard to so many types of sexual activity that there seems to be no sound basis for the widespread opinion that the younger generation has become more active in its socio-sexual contacts."[50] However, in two areas we find changes, which, while not dramatic, are significant.

For one, younger males are showing somewhat higher levels of what the Victorians would have considered "perverse" behavior—masturbation, homosexuality, and petting to climax, especially. For another, there is a general, if modest, rise in most kinds of sexual activity; but intercourse with prostitutes has dropped markedly.[51] This is in part due to educational campaigns about the risks of venereal disease, and in part to the disappearance of the vice districts—although, Kinsey says, the figures "make it doubtful that the number of girls involved in prostitution has been very much decreased."[52] But, taken as a whole, the sexual behavior of these two generations of American males was much the same.

For our purposes, the problem with the Kinsey material on males lies in the fact that his older generation was not truly Victorian, for almost half of them were born after the turn of the century, and were being raised at a time when the new freedoms were already beginning to whistle through the society. These were precisely the people who would as youths insist upon doing the new dances, smoking, drinking. Their sexual behavior, then, cannot be defined as Victorian, because the majority of them were beginning their post-pubertal sex lives in the 1910s, and some of them not until the 1920s. But this older generation of Kinsey's male respondents was being raised in homes that were resolutely Victorian—at least the middle-class portion of it—and we can say that it continued to be colored by the Victorian ethic.

Fortunately, by the time of the female study, Kinsey had seen the value of breaking the figures down into more precise birth cohorts. And it is among the women that we do see a real change in sexual behavior. Interestingly, this change was not manifested by the women who were smoking and dancing and going to night clubs in 1910 and thereafter; it was made by their daughters, or younger sisters. The most significant changes in sexual behavior in the first half of this century were produced by women born between 1900 and 1910. Figures for "petting," a term Kinsey used to include anything from simple kissing to manipulation of the genitals, increased sharply: by age twenty 80 percent of this cohort of women had petted, as against 65 percent of women from the previous generation. Furthermore, the new generation was beginning to pet about a year earlier, and the number who had petted to orgasm by age twenty-five had about doubled.[53]

Premarital intercourse among these women was also sharply on the increase. In the older generation, at age thirty, when most of those women who would marry had done so, about 25 percent had had sex before their wedding days. For the next generation the rate doubled to 50 percent.[54] "This increase in the incidence of premarital coitus, and the similar increase in the incidence of premarital petting, constitute the greatest changes which we have found between patterns of sexual behavior in the older and younger generations of American females,"[55] Kinsey says. The women born in the following decades of the 1910s and 1920s did pet slightly more than did this pioneering generation, but the rates for pre-marital sex actually declined in the next generation. Kinsey suggests 1916 to 1930 as the years when the new sexual morality spread through the social system; these, of course were the years when the women born in the first decade of the century were coming to sexual maturity.[56] Not all women were acting on the new morality, by any means. Half of these women were virgins on their wedding days;[57] and a little over a quarter had had sex only with a single partner, in most cases the men they eventually married; only a small percentage were actually promiscuous.[58]

Nonetheless, this was the cohort of women who made the break with the Victorian past so far as sex was concerned, and the effect on male sexuality was considerable. It is Kinsey's view that, whatever else may have occasioned the lessening male recourse to prostitutes, the principal reason was that the women men knew socially were more willing to engage in sex than they had been in previous generations.[59]

We have seen considerable differences in drinking patterns and other leisure habits between blue- and white-collar classes. Are there similar distinctions between classes in sexual behavior? The Kinsey figures suggest so. For example, by age fifteen some 18 percent of girls who never got beyond grade school, as opposed to one percent of those who would go on to college, had had pre-marital sexual intercourse. In the age group sixteen to twenty, 38 percent of the grade school sample, as opposed to about 18 percent of the college group, had started having pre-marital sex.[60] Similar figures were obtained for males: 98 percent—that is, virtually all—males who stopped going to school before high school had sex before marriage; for those males who started college it was about two-thirds. For those who started high school but did not go on to college, the figure was 84 percent. And the report says, "In the age period between 16 and 20, the grade school group had 7 times as much pre-marital coitus as the college group."[61]

The sexual division along class lines found in the Kinsey sample has to be seen in light of the fact that women from working-class samples were marrying earlier—in some groups much earlier—than middle-class women. In the end, middle-class women, with more pre-marital years in their life histories, would catch up to their working-class sisters. It is also true that, especially among males, non-coital sex, such as masturbation and homoeroticism, were less acceptable in the blue-collar than in the white-collar class. But the Kinsey figures make it abundantly clear that overall, in the first several decades of the twentieth century, blue-collar people had a substantially greater tolerance for sexual activity than was the case in the middle class.

Anecdotal evidence provided by Kathy Peiss, and others, suggests that this had been the case in an earlier day. In her study of working women at the turn of the century she says, "Working-class women received conflicting messages about the virtues of virginity in their daily lives. Injunctions about chastity from parents, church, and school might conflict with the lived experiences of urban labor and leisure." She quotes one working woman as saying, "A girl can have many friends, but when she gets a 'steady' there's only one way to have him and to keep him; I mean to keep him long."[62] Another one said, "Don't yeh know there ain't no feller goin' t'spend coin on yeh for nothin'? Yeh gotta be a good Indian, kid—we all gotta." Another insisted that there was no reason for a girl to "lay down" on the first date, but that it was more acceptable to do so after three or four.[63]

The Peiss study indicates that there was, in at least some working-class groups, an attitude toward sexuality that differed sharply from what was held in middle-class groups. Peiss says, "Sexual knowledge was communicated between married and single women, between the experienced and the naive." In one department at Macy's in New York, "there was enough indecent talk to ruin any girl in her teens who might work on that floor," according to a middle-class observer. Another middle-class observer in a restaurant overheard waitresses "tossing back and forth to each other, apparently in a spirit of good-natured comradeship, the most vile epithets that I had ever heard emerge from the lips of a human being."[64]

Was the middle class now picking up a new attitude toward sex from working people? It is certainly clear that a considerable percentage of middle-class men frequented the vice districts, where they found not only prostitutes, but working-class women in dance halls, casino gardens, and saloons who were looking for men to date, and who might be willing to exchange sex for "treats" and presents. Many of these young women—indeed girls—were attracted to "college boys," and it is clear that some males from the middle class did date these working-class females, did enter into longer term relationships with them on occasion: Edmund Wilson has given in great detail his long affair with a working woman he met in a dance hall in the 1920s.[65]

In this fashion American middle-class males were discovering that the sexual patterns they were being taught at home did not obtain everywhere in the social system: there was, among many working-class females, a much greater willingness to enter into sexual relationships than these males found among the women of their own class, and to this degree it is fair to conclude that, just as the middle class was adapting to its own needs working-class entertainment like the blues and the movies, so it was taking up working-class sexual patterns.

Are we going to say then that the sweeping changes occurring in American attitudes and behaviors that we have been looking at occurred when the middle class began to adopt working-class patterns? Or were there other reasons?

To answer this question—if we can—we must begin with the fact that these changes were happening everywhere in society. Some of these new patterns, like the jazz dances coming out of the dives, were clearly blue collar in origin; others, like the new psychology of Watson and others, were not. Let us look at some.

In particular the year 1912 appears to have been pivotal. I recognize that as sweeping a change to a culture as this one can hardly be placed in a single year. Nonetheless, in the years immediately around 1912— let us say from 1910 to 1915—there was manifested in the society a startling number of parallel phenomena which may have escaped notice at the time, but seem to us today like turning points, or watersheds. Just to mention a few, in 1912 Margaret Sanger published the first of her series of articles on female sexuality for the *New York Call,* a social-ist paper.[66] The Congressional Union, a pivotal feminist group, was founded in 1913, and by 1914 the women had the vote in nine states.[67] *The Masses,* the most famous of all left-wing magazines, was put in the form we know in 1912.[68] The beginning of Greenwich Village radical bohemianism is dated from 1913 by Henry F. May.[69]

The new dances began to spread across the nation in 1909 or so, and the word "jazz" was first mentioned in print in 1913.[70] Sigmund Freud made his famous lectures at Clark University in 1909, and in 1913 "John B. Watson, the father of American behaviorism, proclaimed the revolt of the psychologists."[71] John Chynoweth Burnham says that Freudism and Behaviorism "came into technical and avant-garde literature about 1912."[72] Burnham also dates the rebellion of the intellectuals from 1912.[73] David J. Pivar, writing on the purity crusade, says that in 1912 the Victorian consensus on sexuality was "permanently shattered."[74] The literary critic Julian Symons dates the beginning of literary mod-ernism from 1912, the year in which Harriet Monroe's magazine *Poetry* and the English anthology *Georgian Poetry I* appeared.[75] Robert Wiebe says "Progressivism reached flood tide around 1912."[76] John L. Fell puts the "preliminary institutionalization of Hollywood" in 1913,[77] and Lary May says that by 1914 the middle class had accepted the new medium.[78] The Armory Show, which brought modern art to the Amer-ican public, was held in 1913, and the Buffalo show, which did the same for modern photography, came in 1910.

It was happening everywhere: in art and literature, in philosophy and psychology, in popular culture, in politics, in sex. And it was oc-curring not only in the United States. In England, Virginia Woolf said later, "on or about December, 1910, human character changed."[79] And her sister, Vanessa Bell, reflected, "That autumn of 1910 is to me a time when everything seemed springing to new life—a time when all was a sizzle of excitement, new relationships, new ideas, different and intense emotions all crowding into one's life."[80]

It is critical for us to see that all of these phenomena were coming along together. When we come to look at a similar change that oc-curred in the late 1960s and early 1970s, we will see clearly that a new philosophy put forth by some literary renegades was followed a few years later by a change in attitude, and then in a few more years by a change in behavior.

This was not the case in the years around 1912. For example, the substantial increase in the consumption of alcohol preceded public awareness of the new thought and the new art, but the new sexual permissiveness followed it. The dance boom, in which women played a significant role (we remember Peiss's figures showing that almost three times as many working-class adolescent girls than boys knew the new dances) preceded the feminist movement by several years. We cannot, when looking at these related phenomena, find among them any clear line of cause and effect: they were all coming along together.

What was happening, then? The crucial point, I believe, is that the change in the American psyche occurred precisely at the moment when the swarming city was finally in place. Why? It seems to me that at bottom the industrial city made it *appear* that the Victorian ethic was no longer necessary. People were not living in closely involved families embedded in tightly knit communities. The boss was no longer the father, with whom growing sons and daughters had an intense relationship, but a fairly abstract figure who changed frequently; co-workers were no longer the brothers, sisters, parents, aunts and uncles to whom you were bonded for life, but casual acquaintances who came and went. The family, with its folkways and morality, no longer had the weight that it had, for it could be escaped by even the fairly young for the bulk of each day; and escaped permanently in the mid-teens when most people started to work. And the community—what was it now? Nothing but random, frequently short-term associations. Did the community any longer care for anybody? Did it reform the wayward, solace the grief-stricken, reward the diligent and honorable, heal the sick, rejoice at the birth and bury the dead? It did none of those things. In the old sense of the word the community, the social group, no longer existed.

The middle class, thus, did not acquire the new modes simply through contagion. It was rather that middle-class people, especially but not exclusively the young, were concluding that the Victorian ethic no longer served any purpose; when they reached out for something new, they found out there the movies, the new dances, ragtime and jazz, beer and wine, and the rest of it. And when the middle class shucked off the Victorian ethic in order to adapt a way of life akin to one that had existed for a long time in the working-class culture, American society made a gigantic shift, and what we see today as the modern era, rooted in the industrial city, arrived.

Historians have argued over how the abandonment of the old community should be dated. Some place the change back before the Civil War; others do not see it as having arrived until after World War I; yet others put the moment of change somewhere in between. In my view, it is probably best to see the process as gradual and more-or-less continuous, speeding up here, slowing down there. That is to say, industrialization and urbanization took place here and there in mosaic fashion, gradually filling in across the nation. The seeds of it can be seen as early as 1830 or even before, in the towns growing around the mills of Slater and others; and the industrial city was certainly firmly in place by 1920. But as late as 1870 and even 1880, the majority of Americans still lived in small communities with their tight-knit families embedded in community webs.

The change, in any case, was a shift in human life of very substantial significance. In their public lives people were still required to conform to the exigencies of the work place. But in their private lives, by 1912 it appeared that there were no longer any rules. City-dwellers were discovering that the community did not really care very much what they did—not, certainly, in the way that the old community, with its closely wired human connections, would have. And if there was no community, what was the use of following community values? Why, then, should they not dine and drink and dance as they liked? Who could stop them? Who even cared?

The idea that the actual size of the community can have a drastic effect on the morality inside of it is not simply a hunch. Statistical evidence, taken both today and in the past, supports it. For example, the fight to control, and eventually prohibit, alcohol was to a considerable extent a battle between the cities and the rural areas, and the drys won it because so many of the state legislatures were controlled by rural voters.[81] So far as sex is concerned, right up to today smaller communities follow a more restrictive sexual code than large cities do. A 1985 study of permissiveness and community size concluded that in the biggest cities 30.8 percent of those interviewed accepted extra-marital sex; in the suburbs the percentage who accepted it was down to 19 percent, and in rural areas it was 6 percent.[82] A 1978 study showed that in metropolitan areas 64.8 percent approved of pre-marital intercourse, against 47.4 percent in non-metropolitan areas.[83] A 1970 study by the Kinsey Institute said that "The smaller the community in which the respondents lived, the greater the likelihood they would see homosexuals as abhorrent and dangerous," and added: "The larger the town or city in which people grow up, the less likely they are to worry about being hurt or losing personal dignity in heterosexual relationships; so people with urban childhoods are indirectly more likely to be sexually liberal."[84] And Ira L. Reiss, a highly regarded authority on sexual attitudes says, "In terms of standards, the big cities (of more than 100,000) have an excess of double-standard and permissiveness-with-affection adherents. City size appears to operate as a direct determinant of permissiveness, probably due to its encouragement of courtship autonomy."[85]

All of this is what we would expect. For one thing, it is much easier for people to hide questionable behavior in the more anonymous big cities; and in the smaller, self-contained and more fixed community, a previous misstep may be remembered, and continue to hover, in aura-fashion, over the miscreant.

I suspect, however, that more subtle psychological factors are also at work. It does seem to me that in the small community, where individuals are likely to be more tightly bonded over longer periods of time, where authority figures from the past—fathers, mothers, religious and political leaders—are a continuous presence, the pressures to conform may be greater.

This was seen even at the time. Writing in the 1940s about his return from Europe after World War I, the perceptive Malcolm Cowley said, "We returned to New York, appropriately—to the homeland of the uprooted, where everyone you met came from another town and tried to forget it; where nobody seemed to have parents, or a future beyond the swell party this evening and the disillusioned book he would write tomorrow."[86]

Community size is hardly the only factor affecting attitudes and behavior patterns: religion, age, birth cohort, and other matters enter in. But it is clear that sheer population density has a drastic effect on permissiveness in many areas of behavior, and herein, I think, lies a major key to what was happening to the old Victorian ethic: it was being strangled by the industrial city itself.

It was, then, inevitable that the city became the battleground where the war against Victorianism was fought: for in the end, the city itself was the enemy. And the victor was the self.

CHAPTER 11

The Death of Victorianism

The 1920s have always been seen as one of the most critical decades of modern times, a watershed period during which a new spirit, a new way of thinking, arrived in America. Even as it was happening, people, especially young people, felt that they were taking part in something different, an adventure. They were, they believed, an avant-garde which was towing the future behind them. Edmund Wilson, to emerge as one of the country's most important critics during the 1920s, said, "In repudiating the materialism and the priggishness of the period in which we were born [the 1890s] we thought we should have a free hand to refashion American life as well as to have more fun than our fathers."[1]

Historians, social scientists, began studying the decade even before it was over. Frederick Lewis Allen's famous social history, *Only Yesterday*,[2] appeared in 1931; the Lynds began their even more famous study, *Middletown*,[3] before the decade was even half over; and by 1929 Hoover had established a President's Research Committee to investigate a number of aspects of American culture, which resulted in a spate of books in the early 1930s.[4] (One consequence of all this American soul-searching was to leave a considerable body of contemporary information on the period for the benefit of future investigators.) The Jazz Age, then, was seen by almost everybody then, and by many people today, as an age of innovation, a great creative period in modern literature, art, thought, and behavior. But more recent historians have questioned this view. Was that brief period between the end of World War I and the beginning of the Depression really so innovative, so fresh and daring?

To begin with, the 1920s was a politically conservative period. At its very beginning a United States Attorney General named A. Mitchell Palmer, taking advantage of the fact that President Wilson was desper-

145

ately ill and unable to hold the reins of government, began an attack on "reds" which made the excesses of the McCarthy era of the 1950s look prissy.[5] Palmer jailed people—even deported them—on whim in outright defiance of the Constitution, just because he had decided that they were "bolsheviks," "anarchists," or whatever label he chose to attach.

In time the Red Scare and the Palmer raids simply died out; but they had set the tone for the period: the 1920s would be a time when the ideology of capitalism was a basic tenet of the social system. The old, reforming Progressive spirit, which only a few years before the Palmer raids began was attempting to check the excesses of the industrialists, was dead. "The refrain, let the economy return to natural forces, was repeated endlessly," says David Burner.[6] According to George Soule, during the 1920s "The radical movement growing out of the war had almost disappeared."[7] Paul H. Carter says, "Much that has masqueraded as 'radicalism' in the Twenties was radical only in the sense of *épater les bourgeois*."[8] During the 1920s American voters elected three of the most conservative presidents of the twentieth century: Harding, who turned the government over to business, and presided over the Teapot Dome scandal; Coolidge, who had made his name as governor of Massachusetts by breaking a police strike; and Hoover, a relatively sensible man who nonetheless believed deeply in *laissez-faire* capitalism.

For another thing, the period has usually been seen as one of great prosperity, with the stock market endlessly rising, and a flood of consumer goods, driven by new technology and advertising, coming into American homes—washing machines, electric refrigerators, canned food, and the rest of it. And to an extent this prosperity was real. But it left untouched probably the *majority* of Americans. Farmers were in trouble throughout most of the decade. The farm price index was at 205 in 1920, fell to 116 in 1921 and by 1927 had crept back only to 131. The new science of nutrition helped growers of citrus fruits and green vegetables; but the mass of farmers was hurt during the 1920s and stayed hurt.[9]

Second, the prosperity of the 1920s did little or nothing for blacks. A few in show business, such as Duke Ellington, Bill "Bojangles" Robinson, and some others, benefited from the money available for nightclubbing, but they constituted a tiny elite on top of a vast mass living in deep poverty. Throughout the early decades of the century blacks fled the South, driven out by a wave of lynchings,[10] the introduction of machinery into the cotton fields, and the lure of jobs in northern industry, especially during World War I as armament plants whirled into high gear. The black population of New York went from 92,000 in 1910, to 152,000 in 1920 and 327,000 in 1930; in Chicago it was 44,000 in 1910, 109,000 in 1920, and 233,000 in 1930. In other cities, such as Detroit, the percentage of blacks in the population rose even faster.[11]

The cities, however, were already bursting at the seams with the immigrants and the influx from the farms and could—or would—do little for blacks. They were crammed into their own enclaves, like the Black Belt in Chicago's South Side or New York's Harlem, where landlords cut apartments into small units and jacked up rents. In order to pay the rent many, if not most, black families were forced to take in lodgers, as so many immigrants had done. Said one authority at the time, "In many blocks of these cities there are more lodgers in the households than children." The same contemporary observer said that black sections of the big cities were the "most densely populated sections of America. Blocks with from 250 to 350 people per acre are not uncommon." This density was something like four times that of white areas in some cities.[12] "Municipal services were of lower grade in Negro sections, paving and sewerage were often obsolete, and sanitary inspection less adequate,"[13] the same authority said. The Harlem to which intellectuals, slummers, and big spenders from out of town visited to see the show at the Cotton Club, the Nest, or Connie's Inn was by daylight a fetid slum filled with bitterly poor people trying simply to get from one day to the next.

Third, the working class, which by itself was the majority of Americans, did not get a fair share of the prosperity of the times. Sixty percent of Americans were at or below subsistence level,[14] a figure that seems more reasonable when we remember that it included virtually all blacks, probably the majority of farmers, and a substantial proportion of recent immigrants. The Lynds flesh the figure out with a picture of the ordinary working family with little or no savings, no unemployment insurance, always on the margins, living in constant fear of the layoff which inevitably came, and facing a future in which by the age of forty-five or so the men would be unemployable, except in the most menial of jobs.[15] The explanation for this very unequal division of the spoils during the 1920s is quite simple: from 1900 to the end of the 1920s productivity rose 50 percent; but during the same period the standard of living for labor rose only 25 percent.[16] The middle class, and the rich, ensconced in the seats where decisions about wages and salaries were made, were paying themselves decent incomes, but were driving wages on the shop floor down by way of the competition of men for an insufficient number of jobs, while a business-oriented national government gazed around benignly.

The prosperity of the 1920s belonged to the middle class, and they used it for their own enjoyment. They were able to do this at least in part because many of them found themselves with substantial amounts of leisure time on their hands. The twelve-hour day, six-day work week was gone. Between 1900 and 1930, hours of work declined by 15 percent.[17] The ten-hour day and the half-holiday on Saturday were coming to be standard. The free Saturday afternoon gave them time to do routine chores so that Sunday might be largely devoted to pleasure.

The young especially were gaining free time. High school enrollments increased eightfold from 1900 into the twenties, and college enrollments fivefold.[18] Many of these high school students, who were mainly from the middle class, worked at part-time jobs after school and on Saturday, but large numbers of them did not. In 1900, 62 percent of males aged fourteen to nineteen were in the work force; in 1920 it was 51.5 percent.[19] Under the influence of the ideas of Dewey and his followers, schools were attempting to educate the whole human being, and as a consequence were adding to the basic curriculum courses in art, crafts, homemaking, social activities. Schools began to sponsor drama clubs, newspapers, dances, concerts, and the like. These frequently took place after school and many parents, especially in the middle class, felt that it was more important for their children to participate in these activities than to hold after-school jobs.[20] "Today the school is becoming not a place to which children go from their homes for a few hours daily, but a place from which they go home to eat and sleep,"[21] the Lynds said. The young, especially of the middle class, now constituted a semi-leisured group with a great deal of disposable time.

Women, again especially those in the middle class, were claiming even greater numbers of disposable hours. Very few of them worked outside of the home after they married: by 1930 only one of eight married women was employed, and that includes women of the working class as well.[22] A considerable proportion of the middle-class women had servants: a third had full-time help, and 90 percent had some sort of part-time help, if only a cleaning woman or laundress who came in one day a week.[23] Moreover, the middle class could afford to buy the new labor-saving devices coming on the market, and were increasingly feeding their families with prepared food. Sales of baked goods, especially bread, were up 60 percent in the years from 1914 to 1924. By 1920 there were three times as many delicatessens in the United States as there had been ten years before. The use of commercial laundries increased by 57 percent between 1914 and 1924.[24] The consequence of all of this was that by the mid-1920s there had come into existence a "large group of [wives] who by careful management fit everything somehow into the morning and an afternoon hour or two and contrived to keep many afternoons and evenings relatively free for children, social life, and civic activities."[25]

Working-class women also, to an extent, benefited from some of the same advantages that were giving middle-class women so much free time. Most of them did not work after marriage, and, as with middle-class women, fewer children made for less work at home. By the time they were in their thirties, most of their children were in school for the bulk of the day, and then in jobs and establishing homes of their own. These women, too, were buying bakery bread, canned peas and beans, rather than baking their own bread and cakes, shelling their own peas, and stringing their own beans; they, too, were managing to buy refrigerators and washing machines; they, too, were taking their sheets and their husbands' shirts to the commercial laundry.

The existence in America of a large class of women with a lot of spare time at their disposal was something brand new, which had come into existence, really, only after 1900, and particularly after the end of World War I. There had, of course, always been a class of idle rich women with servants and little to do but amuse themselves, but this group had been very small, perhaps not more than a percentage point or two of the population. Prior to, let us say, 1890, most women were occupied full time by the demands of their families. Farm women routinely worked a ten- to twelve-hour day,[26] much of that time doing hard physical labor like churning butter, milking cows, scrubbing floors, and washing clothes in large tubs. But by the 1920s, through a combination of many things, possibly a quarter of American wives and their daughters had many hours in each day to do with as they chose.

Men, too, were gaining more leisure time. Not only was the work week shorter, but a new invention, the paid vacation, was becoming standard. In the 1890s the vacation, even for high level executives in business, was simply unheard of: everybody worked fifty-two weeks a year, with time off only for the standard holidays—July 4th, Christmas, Thanksgiving, Memorial Day. By the 1920s the paid vacation, a week or two long, "is increasingly common" for white-collar workers.[27] For working people it was still rare. Nonetheless, in Middletown some plants gave foremen a week or so off with pay, and in many factories a worker could take a week off without pay, if he chose. Other plants sometimes simply closed their doors for a week or two in the summer, giving their workers vacations without pay, whether they wanted them or not.[28] The paid vacation was by no means universal in the 1920s, but it was nonetheless a fact of life for millions of Americans.

The effect of the paid vacation on the economy was enormous, for it gave rise to the huge vacation industry that exists today. By the end of the 1920s the American Automobile Association was reporting that some 40 million people were making automobile vacation tours each year.[29] The American bill for foreign travel, at a time when this meant a long journey by ship and therefore usually required more than two weeks away from work, increased from $356 million annually to $898 million.[30] The summer home was now a possibility, and all over America bungalow colonies began to sprout along the shores of lakes, rivers, and oceans.

This massive new leisure time, which in the aggregate ran to—very roughly—something like two billion man hours *a week,* could not avoid having enormous consequences for the society. And it was coupled with one other phenomenon with even greater effects: the sudden emergence in the 1920s of the automobile as a commonplace in American life. A car, before 1900, was for most people a strange and astonishing sight. According to the Lynds, not until that year was there a real car in Middletown, and as late as 1909, when the automotive industry was coming into its own, there were manufactured in the United States two million horse-drawn carriages.[31] But during the 1910s the car was becoming more common. Then, in World War I millions of American soldiers became familiar with trucks, by driving them, and certainly by riding in them. By 1920 there were 9.2 million cars in the country; by 1925 the number was almost 20 million and by 1930 it was 26.5 million.[32] In 1923 about two-thirds of Middletown's families had cars.[33] In another small town studied by social scientists, "The car became possible and popular in Eno Mills about 1914. In that year the number of cars increased from five or six to twenty or twenty-five, and within five years two-thirds of the families in town were car-owners."[34] The working class was not left out. Working families told researchers that they would rather cut down on food, clothing, even give up home ownership in some cases, in order to own a car.[35] In a number of instances laboring families had no bath tubs but had automobiles. In the 1920s the automobiles became "an accepted essential of normal living."[36]

The consequences of widespread car ownership were large. Previously, working people especially had to walk to work, and therefore had to find homes near the factory, or jobs near their homes. The result was a clustering of people working in a given mill around the factory, which inevitably gave the neighborhood a certain cohesiveness: layoffs, strikes, job problems were the common property of the people in the local saloon or the grocery store. But as working men bought cars they found it worthwhile to look for jobs at greater distances from home, where the pay, or the conditions, might be better, and neighborhoods began to lose their cohesiveness—yet one more force breaking down the old idea of community.

But perhaps more significant was the way that the automobile worked in tandem with the new leisure time, for it presented people with new ways to expend the extra hours. For one thing, one of the major recreational activities of the 1920s was the Sunday automobile ride. It might be a visit to relatives who previously would have been seen mainly on major holidays; it might be a trip to a lake, a mountain, an ocean beach some miles distant, which before had been only a name and a picture on a postcard; it might be a spin in the countryside to see the autumn colors, or simply to escape the noisome city. In the 1920s the car was for most people a new, liberating toy which excited them, and they drove for pleasure almost as much as out of necessity. And to this extent it brought families together, for it was quite customary for parents and children to pile in the car on Sunday afternoon and go for a drive.

But the more general tendency was in the opposite direction. What the car allowed people to do was to pursue activities which they might otherwise have not been able to. A car made it possible to get to golf courses, ski resorts, distant beaches, woodland parks. Country clubs were now closer. It became easier for people living in Brooklyn, or Chicago's Northside, to get to the jazz clubs of Harlem, the South Side, Greenwich Village. Wives could travel considerable distances to meet friends for lunch, tea, bridge, or less seemly activities. The automobile touched on almost everything.

The young in particular seized upon the automobile and made it central to their lives. The Lynds found that car ownership was a social necessity. "Among the high school set, ownership of a car by one's family has become an important criterion of social fitness: a boy almost never takes a girl to a dance except in a car; there are persistent rumors of buying of a car by local families to help their children's social standing in the high school."[37]

But the car had, for the young, more than a status value. Its primary function was to allow adolescents, who were still living at home, and theoretically under the rule of their parents, to escape authority. In an earlier time, when young people traveled by foot, or at best by streetcar, they were almost forced to socialize in neighborhood groups. The tendency, in an earlier day, was for the young to operate in packs, getting up dances, picnics, skating parties among themselves, in which they did not necessarily pair off. The car had limited seating space—we must remember that the inexpensive automobiles most ordinary families bought at the time usually carried only four people comfortably, and if it was a roadster with a rumble seat, two of them were not even comfortable, especially on a cold night. Pairing off became more common. In Middletown the Lynds recognized a "growing tendency to engage in leisure-time pursuits by couples rather than in crowds, the unattached man or woman being more 'out of it' in the highly organized pair social life of today than a generation ago when informal 'dropping in' was the rule."[38]

The automobile was not by itself responsible for the increased tendency to pair off. The growing acceptance of petting, or necking as it came to be called, tended in the same direction, and undoubtedly other factors were at work, all reinforcing the others. But through the 1920s there was a broad movement away from the touring car, which afforded passengers clear views and the smell of fresh air, to the enclosed sedan: in 1919 only 10 percent of cars were sedans; by 1924 it was 43 percent; and in 1927 it was 82.8 percent.[39] The car was no longer a vehicle for sightseeing, but a portable living room, and could be used for eating, drinking, smoking, gossiping, and sex. The Lynds noted that in Middletown, of a group of thirty girls charged with "sex crimes," for nineteen of them it had happened in a car.[40]

The use of the family car not only gave the young a private place in which to carry on private matters, but it allowed them to make forays to distant places miles from the authority of parents, teachers, and chaperons—secluded byways, roadhouses, dance halls, and the like. Once they were in the car they were free; and they took every opportunity to get away that they could: over half of Middletown's young were away from home four or more evenings a week.[41]

The effect of the car on the family, then, was centrifugal. "The family is declining as a unit of leisure time pursuits,"[42] the Lynds said flatly. Once again, the car alone did not produce a fragmenting of the family; but it made it a lot easier.

It should be clear, then, that the primary use to which Americans put both the automobile and the new leisure was recreation. The bare statistics tell an astonishing story. Half of the 25,000 tennis courts in the United States at the end of the 1920s were built in that decade.[43] The number of golf courses went from 742 in 1916, to 1,903 in 1923, to 5,857 in 1930.[44] Public golf courses soared from 24 in 1910 to 543 in 1931.[45] Between 1927 and 1929 alone the value of golfing equipment rose 71.8 percent.[46] Again, in 1920 there were 25 ski clubs in the United States; in 1930 there were 110.[47] The number of bowling alleys doubled in the 1920s.[48] In 1924 161 cities counted 74,000 softball players in numerous leagues; by 1930, some 344 cities reported 213,000 formally organized players.[49] The number of hunting licenses issued annually in the decade increased by 69 percent, and the value of fishing equipment bought increased 13 percent in the last three years of the decade alone.[50] Cities with public bathing beaches increased from 127 to 218 in the years between 1923 and 1930; Chicago reported doubling of beach use from 1925 to 1930.[51] The number of swimming pools rose by 80 percent from 1923 to 1930,[52] and the registration of motor boats went from 130,000 to 248,000 during the decade.[53] By 1930 docking facilities were "entirely inadequate," and many cities reported serious water pollution problems.[54]

The coming of the automobile spurred a boom in the use of parks and campsites. In the decade of the 1920s the number of state parks tripled.[55] Visitors to the great national parks skyrocketed from 334,000 in 1915 to over 3 million in 1931.[56] Visitors to the national forests soared from 4.8 million to 31.9 million in the decade.[57]

The number of urban parks, too, grew furiously. Between 1907 and 1930 city park acreage zoomed from 76,000 to 258,000, while the urban population rose by 65 percent.[58] Furthermore, there was a change in the nature of parks: whereas in the 19th century they had been meant for the "quiet enjoyment of well-landscaped, wooded places,"[59] they were now being turned into entertainment centers. "In a remarkably short period of time urban parks from one end of the country to the other have been equipped with a bewildering array of leisure-time facilities designed to meet the needs of both young and old," among them bandstands, field houses, outdoor theaters, picnic tables, croquet grounds, softball fields, outdoor cooking facilities, lakes for boating and much more.[60] By 1930 cities had invested two billion dollars in their parks.[61]

But if people were playing more, they were also watching others play in growing numbers. Seating capacity at college football stadiums more than tripled during the 1920s.[62] Boxing, which had hitherto been anathematized—indeed forbidden by law at times and places—grew rapidly in popularity, in part because many American men had learned to box as part of their army training during World War I. Receipts for the most popular bouts averaged around $60,000 in the first decade of the century; about $200,000 in the second decade; and a million dollars in the 1920s.[63] It is significant that only professional baseball saw little growth during what has frequently been called the Golden Age of sport. During the decade admissions increased by about 10 percent, while attendance at minor league games actually declined, because there was so much more for Americans to do now. The report said, "Baseball is approaching the peak of its popularity."[64] (That was a bad guess: by 1987 baseball attendance was over 50 million.[65])

Movies, which had brought in the middle class in the previous decade, were now a national phenomenon and continued to grow. In 1922, 40 million Americans were going to the movies on a weekly basis; by 1928 it was 60 million.[66] That was the *majority* of the American people. In some places they were going even more often: in Middletown in July 1923, the month with the lowest movie attendance, three-fourths of the city's population went to at least one film. A survey of Middletown's high school students revealed that only about a third of them had not gone to a movie in the previous seven days, and about another third had gone twice or more in the same period.[67]

In the 1920s, too, the phonograph truly came into its own. Technical advances, especially the introduction of electrical recording in the middle of the decade, greatly improved sound quality. The use of the microphone in place of the old acoustical horn permitted companies to much more easily record large groups, like symphony orchestras. By 1921 record sales were up to $100 million annually, four times what they had been in 1914.[69] It was not, however, a growing interest in classical music which fueled the booming market for phonograph records. The key figure was a fat, genial, former symphonic viola player named Paul Whiteman, who led a band playing jazz-based dance music. The new sound had been concocted mainly by another symphonically trained musician and composer, Ferde Grofé, and band leader Art Hickman, for whom Grofé probably made the first modern dance band arrangements. Whiteman hired Grofé as pianist and arranger in 1919.[70] The group's coupling of "Whispering" and "Japanese Sandman" sold 2.5 million copies, and its "Three O'Clock in the Morning" sold even more[71]—enormous sales at a time when the population of the country was half of what it is today, and records cost the equivalent of two hours' work for an ordinary laboring man.

Whiteman's huge success brought dozens of other band leaders in his train, and it was this new jazzy dance music that drove the phonograph industry to ever higher grosses. So great was the demand for top quality dance orchestras that salaries for musicians were forced up to levels which will make present-day musicians howl with envy. Ross Gorman, Whiteman's star clarinetist, was earning an astounding $400[72] a week at a time when you could buy a new car for $525. There appeared to be no limit to the public's appetite for dance music. The bands were everywhere, and by the end of the decade the top band leaders were among the country's most celebrated figures. Paul Whiteman, the king of them all, met with governors and senators as he traveled. As with the movies, the public appetite for popular music seemed insatiable.

Where people found the time to read is puzzling, but they did. Magazine circulation, which was under six million copies in 1900, rose to 16 million in 1910, 22 million in 1920, and 33 million in 1930.[73] In Middletown 90 percent of middle-class families subscribed to three or more magazines, and 60 percent of blue-collar workers regularly read at least one.[74] In 1923 almost half of the people in Middletown in 1924 had library cards, more than double the figure for 1910, and were taking out 6.5 books per resident each year.[75]

Drinking, of course, is not nearly so time-consuming as reading a book or listening to a phonograph record, if for no other reason than it can be done in tandem with many other activities. How much drinking was done during the 1920s is a matter of dispute. Prohibition was, if not exactly in force, at least on the law books throughout the decade. Until fairly recently it had been taken for granted that drinking rates rose under Prohibition. People who had never drunk much before, it seemed, began visiting speakeasies, or slipping around to the local bootlegger for a bottle "right off the boat." It was perfectly obvious that a lot of women, who in a previous time would not have considered going into a saloon, were now frequenting speakeasies, and openly drinking whiskey from pocket flasks at football games. The view has been that Prohibition, rather than prohibiting drinking, had actually popularized alcohol as part of the swing against the Victorian ethic.

The idea has been hard to challenge; reliable statistics are difficult to get because there were no state or federal taxes collected on the sales of bootleg liquor. But careful studies by Joseph R. Gusfield, Norman H. Clark,[76] and others suggest that, if anything, the amount of alcohol drunk during Prohibition actually fell. As we have seen, drinking reached a peak during the early years of the century, despite the force of the prohibition movement, and it is probable that a broad rise in drinking above those levels was not really possible, unless the country reverted to the drinking habits of the 18th century when many people drank at frequent intervals throughout the day.

It is important to recognize that many people, perhaps the majority of Americans, believed that prohibition was a good thing, and were ready to give up alcohol as a price worth paying for the end of the saloon, the drunken wife-beating husband, the dissolute criminal stealing to pay for liquor. By 1919, when it was clear that national Prohibition was coming, it was naively believed by a large part of the population that it would work.[77] Of course there were a good many people who felt that there could be too much of a good thing, and did not want the country to go absolutely dry. The rich and even the moderately wealthy stocked up, and many clubs put in cellars of liquor.[78] But the bulk of Americans could not afford to put away anything more than a bottle or two and were resigned to doing without.

Once Prohibition was in place, such liquor as was at first available was priced beyond ordinary pocketbooks. Before Prohibition the price of a drink at a first-class bar was about fifteen cents. The minute Prohibition took effect the price even at a cheap saloon jumped by from forty to sixty cents, and much more in the better places,[79] when twenty-five dollars a week was a good wage for a working man. The immediate effect was to sharply curtail the drinking of working people, who constituted some two-thirds of the American population. The effect alone would have reduced the country's total alcohol intake substantially. The incidence of the diseases of alcoholism was down markedly,[80] and there was a "dramatic decline in the number of arrests for public drunkenness."[81] Estimates by various authorities suggest that consumption of alcohol during Prohibition decreased by a third to a half.[82]

But, as time passed, the illegal alcohol industry began to get itself organized, and liquor at more reasonable prices came on the market. Indeed, the very morning of the first day of Prohibition federal agents seized truckloads of illegal liquor, and raided stills in a number of places.[83] As early as 1919 a Chicago lawyer named George Remus was founding a business in illegal alcohol,[84] and by the mid-1920s the big time gang syndicates had taken control of the business and organized it, along the way paying countless millions of dollars in bribes and earning profits for themselves that are probably inestimable, but certainly ran into the billions of dollars. Nearly everybody over the age of six knew where liquor could be bought, and the numbers of people who were prepared to obey the liquor law had dropped and continued to fall. As early as 1922 a *Literary Digest* survey concluded that 61.4 percent favored the modification of the Volstead Act, generally to permit the drinking of beer and wine at home; by 1926 another poll showed that 81.1 percent favored modification of the Act or outright repeal.[85] It was perfectly obvious that Prohibition not only did not prohibit any-

thing, but had created a monstrous criminal empire which was cor-
rupting huge sections of the law-enforcement system in the United States.

Nonetheless, obtaining drinks was against the law, and occasioned
certain awkwardnesses, risks, and expenses. Although drinking rates
probably began to rise after 1923,[86] it is doubtful that at any time during Prohibition did they begin to approach the rates that had held in
the previous era. In 1934, the first year in which alcohol was again
legal, per capita intake of absolute alcohol was .97 gallons,[87] as compared with the 2.6 figure for some pre-Prohibition years, although of
course this figure was undoubtedly skewed by the fact that the liquor
industry was only beginning to get up to speed.

Yet this drop in alcohol intake masks a dramatic shift in American
drinking behavior. The prohibition movement had been driven by the
middle class, and openly opposed by most of the working class, especially the immigrants who made up so large a proportion of the industrial labor force. Working men had had their saloons, but probably the
majority of middle-class homes before 1920 did not stock alcohol, and
did not serve it except on ceremonial occasions, like weddings.

Prohibition reversed this class-drawn system. Working people, de-
prived of their saloons, and unable to afford any large quantity of liquor, were becoming relatively abstemious. Now it was the middle class's
turn to drink. It is very hard to produce anything but impressionistic
evidence for this statement, but it certainly has been what both contemporary observers and later analysts have concluded. For example, in
1918 there were 8,168 licensed saloons in New York City; by the end
of the 1920s there were "tens of thousands" of speakeasies. One estimate puts the figure at 32,000.[88] Prohibition, says Joseph Gusfield, "may
well have increased the hard and excessive drinking among precisely
those groups that had in the past been pace-setters and style-setters."[89]
And he adds, "Certainly all the surveys have indicated that middle and
upper classes represent higher levels of drinking than working and lower
classes," ever since.[90]

It is critical that a large percentage of these new middle-class drinkers were women. Again, we have no survey evidence for the rise in the
numbers of women who regularly drank; but an examination of drawings, cartoons, and advertising illustrations in about one hundred issues
of the *New Yorker* and *Vanity Fair* magazines for the late 1920s, magazines aimed at relatively sophisticated audiences and therefore more
willing to acknowledge middle-class drinking than the mass audience
magazines, shows women in almost all depictions of drinking situations,
usually in roughly equal numbers to men, paired off in couples.[91] It
was the perception of the editors of these magazines and their readers
that middle-class women were routinely drinking along with men. Certainly the fiction of the period indicates the same thing, as for example

the many drinking scenes in the writing of F. Scott Fitzgerald, widely read by the middle class of the time.

We are seeing yet again the adoption by women, and especially middle-class women, of a pattern of behavior which a generation earlier had been reserved for men. The fact that middle-class women were drinking meant that men could drink in circumstances where they had not before, in particular in their own homes. The cocktail before dinner, wine with the meal, the scotch and soda over the bridge table, were now permissible; and the very fact that people could drink at home made the speakeasy, the bar, less necessary.

Legal restraints on alcohol limited consumption, but there were no such constraints on smoking. Tobacco was, almost from the opening up of the American colonies, of major importance to the country's economy, and its use was taken for granted.[92] Through the latter part of the 18th and much of the 19th century there was a trend away from smoking tobacco to chewing it. This was due in part to the Victorian distaste for tobacco in any form; according to Jack J. Gottsegen, in his study of tobacco consumption, per capita use dropped from 7.43 pounds in 1790 to 5.65 pounds in 1839, as the Victorian age was coming on, to 1.78 pounds by 1871.[93] But in about 1870, as the industrial city was taking shape, there began a marked rise in cigarette consumption. In the 1880s American manufacturers were producing about 1.3 billion cigarettes annually; twenty years later the figure had tripled; and during the 1920s production rose from an average of 80 billion cigarettes annually to 141.4 billion.

Cigarettes were being exported, of course; but a huge percentage of them were being smoked at home. By the latter years of the 1920s per capita cigarette consumption had reached 1,285—64.5 packs for every man, woman and child in the country. Ten years later the figure had risen by 38 percent.[94]

As with drinking and smoking, so it was with sex in the 1920s: the revolution in sexual attitudes and behavior that was becoming visible in about 1916, especially in bohemian, feminist, and intellectual circles, continued to leach through the society, as the new freedom was accepted by more and more people, especially in the middle class.[95] The Kinsey figures for premarital intercourse, and especially petting, continued to rise.[96] This last may have been the most significant change in sexual behavior to occur in the 1920s. In the generation born before 1900 about 65 percent had engaged in premarital petting. For the next cohort, which was coming to sexual maturity in the late teens and through most of the 1920s, the figure was 80 percent. This was the generation that made the sexual revolution; and it shows particularly in the numbers who were engaging in petting to climax: in the ten years separating the older generation from the newer one, the numbers of women who had petted to climax rose by 70 percent, a dramatic shift in behavior for so short a period of time.[97]

We are not surprised to find, then, that by the mid-1920s the high school students had institutionalized the "petting party." In Middletown, 88 percent of boys, and 78 percent of girls had attended such parties,[98] and this, mind you, was not a wealthy suburb of New York City, but Muncie, Indiana, in the heart of the Midwest.

From this point onward, right through the 1920s and beyond, there are steady increases in sexual activity outside of marriage on almost all fronts. It was a time of consolidation, when the new sexual attitudes and accompanying behavior spread through the social system to become the broadly accepted American way. Not officially: the laws still forbade sex outside of marriage; the churches still termed it a sin; institutions like the settlement houses and the YMCA fought against it; and most parents did what they could to keep their daughters from engaging in most kinds of sexual activity. But beneath the official morality lay the real mores of America, which permitted a great deal more sexual activity than was officially tolerated.

Inevitably, this new view of sexuality manifested itself in dress, especially the costumes of women. An examination of issues of magazines[99] for the period show skirts just above the ankle in 1916, well above the ankle but below the calf by 1918, at mid-calf by 1920, and by 1927 just covering the knee, which was revealed when the legs were crossed. However, these were magazine illustrations and advertisements, and were generally conservative. Women usually bought their dresses a size shorter than what was recommended, and in fact by 1920 skirts were nine inches from the ground, which for most women would put them well up the calf.[100] By the end of the decade skirts when hanging naturally were at or above the knee.

The use of cosmetics simply went through the ceiling in the 1920s. In the older world cosmetics had been limited to discreet touches of powder or rouge, and among the gentry were not used at all. Now that was all done with. Between 1914 and 1925 sales of cosmetics rose from $17 million to $141 million annually. In the 1920s it was estimated that 71 percent of American women over eighteen used perfume, 90 percent used powder, 73 percent toilet water, 55 percent rouge.[101] The whole notion of what it meant to be a woman had changed. Females, even quite young ones, were no longer expected to appear modest and virginal; they were instead to look experienced and worldly-wise. They were to be "men's casual and lighthearted companions . . . irresponsible playmates . . . hard boiled adolescents," according to Frederick Lewis Allen.[102] The slim, boyish figure with small breasts and narrow hips was now the ideal. Bobbed hair was nearly universal among younger women, and was being adopted by many older women, too. The double standard had vanished: not only were women to drink, smoke, and have sex, like men; they were to look like men, and indeed think like men as well. Behavior patterns which only a generation or so earlier would have marked a woman a prostitute were now being adopted by the teen-aged daughters of middle-class families; costumes that would once have only been seen in brothels were now being worn to church.

To an extent, World War I furthered these changes. It carried millions of men to France, along with a lesser, but not insignificant number of women, where they were confronted by a culture markedly different from the one still climbing out of Victorianism they had left at home. It was a culture in which wine was served with meals, even breakfast in some cases; where brothels were legal and taken for granted. The Kinsey Report concluded, "Never before had so many Americans come into contact with a foreign culture, and especially with one in which the sexual patterns differ as much as Central European patterns do from our own."[103] (This, of course, was written in about 1950.) Frederick Lewis Allen said that among the Americans who had been sent to France "there had been a very widespread and very natural breakdown of traditional restraints and taboos."[104]

But although the war, as war often does, helped to break down restraints in a number of areas, the movement away from Victorianism had begun a decade earlier. It was already under way, relentless and unstoppable; the wartime experiences of many men and women may have given it an additional thrust, but it would have continued of its own momentum without any outside help.

One consequence of all of this was a declining interest in religion. Church and Sunday school enrollments failed to keep pace with the population growth. Not more than 60 percent of people attended church regularly in most places, and in the West the figure was closer to 25 percent.[105] A YMCA survey of the members of the American Expeditionary Forces in France concluded that "America is not a Christian nation in any strictly religious sense," and another report said, "Extrapolations of trends suggest the probably further decline of interest and belief in traditional Christianity."[106] This turning away from the churches was probably inevitable: most religions were still speaking against drunkenness, sex out of marriage, smoking, too much nudity, and even such activities as dancing, Sunday sports, and jazz music. The people were headed in another direction, and they did not want to cart their Sunday morning hangovers to church only to be told they were sinners. They knew that, for they wanted to be sinners, and they believed that the cure for their excesses was not prayer, but the lunchtime Tom Collins.

Looked at as a whole, it is clear that the 1920s was the time when the new freedoms, developed earlier by an avant-garde, spread through the culture as a whole. It was the period when the middle class began to drink, when women began to smoke, have sex, and dress in reveal-

ing clothes; when families were splintering, women and children becoming more independent, and everybody was gaining more leisure
time; when the appetite for play ballooned at an astonishing rate, and
the construction of recreational facilities became a large industry; when
the automobile became a major element in the culture; when for the
first time the majority of the population was devoted to movies, popular music, dancing, and the entertainment industry in general. The
American people had become preoccupied by the idea of "having fun,"
whatever they took that to mean. Said one contemporary observer, "The
entire weekend has for large numbers of people been turned entirely
over to the pursuit of pleasure." [107] This is not an idea that would cause
any furrowed brows today; but in the 1920s it was a new, and to many
older people, shocking idea. Jesse Frederick Steiner, who studied
American leisure habits for President Hoover's Research Committee at
the close of the 1920s said:

> During recent years people have become much more concerned with the
> recreational side of life and insist far more than in the past upon easy
> access to sports, amusements, and other leisure time diversions of a widely
> varied nature. While recreation has always been a matter of deep human
> interest, it now occupies a more fully accepted position in the scheme of
> human affairs and finds ready justification on the grounds of health and
> efficiency as well as relief from daily toil. In a very real sense recreation
> has forged to the front as one of the compelling interests in human life
> and has already developed to the point where it makes extraordinary de-
> mands on the time and energy and requires large financial expenditures
> to cover its mounting costs. [108]

As had been the case in the increasing levels of drinking, smoking,
and sex, the adoption by women of the new mode was an important
factor in its widespread acceptance. Elaine Tyler May, in her study of
divorce for the period, discovered that a significant complaint of husbands was that their wives were neglecting their homes in order to have
fun. Court records "suggest that women may have been more attracted
to the new urban amusements than men . . . More and more women
were participating in leisure endeavors, often at the expense of what
many husbands felt were their domestic responsibilities." [109]

The point of course was that virtually all men worked at jobs that
called for their presence at specified hours, where women with children
at school and convenience products in the stores were more and more
able to carve out leisure time for themselves to have lunch with friends,
play bridge, or go to a movie in the afternoon. And if housewives could
permit themselves to amuse themselves during the daylight hours,
something unthinkable in their mothers' time, why should not everyone else?

The 1920s did not invent jazz, the new dances, the concept of sexual freedom, a predilection for drink and the movie matinee. " 'Lost' or otherwise, the generation of the Twenties had inherited a larger stock of ideas from the Gilded Age than it cared to admit,"[110] says Paul A. Carter. Gilman M. Ostrander adds, "By the outset of the twenties the revolution in morals was already completed as far as many American intellectuals were concerned."[111] What happened in the 1920s was that the attitudes and behaviors developed by the avant-garde in their Bohemias spread through the population at large: what had been done by the few was now being done by the many. The 1920s were, as John Chynoweth Burnham puts it, a "watershed." He says that Americans of the 1920s had become caught up in a "cult of the self . . . In a remarkable reversal from an earlier day, social norms produced not only self-centered attitudes but self-indulgent behavior."[112]

The Victorian Age was finally over. It lingered, of course, in the homes of older people who had been raised in the Victorian manner, and to a decreasing extent in the lives of many of their children who did not entirely escape the Victorian hand in their own upbringings. Virginity before marriage, for females at least, remained an ideal into the 1950s, and was adhered to by a considerable proportion of the population. Many homes would not tolerate alcohol, and excessive drinking was generally frowned upon, as was smoking for the young. But effectively by the end of the 1920s Victorianism was dead. In its place was a new ideal of "instant, rather than delayed, gratification; indulgences and excesses rather than loyalties and responsibilities; impulse, rather than duties; the peer-directed, rather than the conscience-driven comportment," says Norman H. Clark.[113] The ideas of 1912 which were aimed at producing a freer, more human social system were, by the end of the 1920s, being used to justify a nationwide binge of self-indulgence.

CHAPTER 12

Hard Times

The story of the great stock market crash of 1929 and the long, agonizing Depression that followed has been told many times, and it is only necessary for me to sketch it in briefly here.

Through the early 1920s the course of the stock market was generally up. There were "breaks" in the line, often quite sharp ones; but stock prices would always go up again. By 1927 there were at least a few people who were alarmed by the heights to which stocks had risen, and some were saying that stock values were inflated well beyond what they were worth by any of the ordinary measures. But the few who felt uneasy were vastly outnumbered by the many who were beginning to see Wall Street as a kind of perpetual motion machine that would allow anyone who knew how the game was played to get rich. What triggered the great Bull Market that began in 1927 has been debated at length, but key to it was a wave of breezy optimism that impelled huge numbers of otherwise sane people to believe that money could be manufactured out of numbers. Americans had, as John Kenneth Galbraith put it, "a mass escape into make believe,"[1] and were being egged on, according to the historian Ellis W. Hawley, by "high pressure securities merchants and promoters with exaggerated and irrational visions of the future."[2]

Through 1928 and into 1929 the Bull Market roared on. To be sure, there were occasional breaks, some of them quite sharp, which ruined a few investors, but each time the market righted itself and started on up again. The value of certain favored stocks doubled, and doubled again. Stock brokers and their banks lent and lent again to allow their customers to buy "on margin," as if nobody would ever have to be paid back. Five-million-share days, unheard of a year or so earlier, became common.[3]

September 3, 1929, was the high point. By this time stocks were vastly overvalued and a number of people had begun to grow wary. Apparently during the long Labor Day weekend, and the vacations running up to it, some speculators had reflected, and decided to pull in their horns. For the next month or so the market wobbled slowly downward, and by the last days of October a sense of uncertainty was setting in. On October 23 the market broke abruptly, with the *New York Times* industrial index losing 18.24 points.[4] Finally the well of optimism had run dry, and the next day the bottom fell out. In one day many stocks lost a third of their value, and almost 13 million shares changed hands.[5] On Monday, October 28, stocks fell with a great crash; and on October 29 what remained of the dam gave way. The *New York Times* was off by almost 40 points. Down, down, the rabbit hole went Alice, until November 13 when a bottom—at least for the moment—was reached. Montgomery Ward, one of the great favorites of the boom, had dropped from 400 to 49½.[6] But although the market showed occasional signs of reviving, it continued its downward trend into the middle of 1932. The *New York Times* industrial index, which had dropped to 224 by November 13th, was now at 58.[7]

What caused the great stock market crash of 1929 is once again something historians argue about. There is, however, fairly general agreement that there was a very considerable weakness in the economy. According to Hawley, there was in fact a great deal of real prosperity in America during the years 1923 to 1928. The era "came close to combining full employment and rising living standards with stable prices and international peace."[8] And he adds, "Taken as a whole, the industrial statistics for the 1920s tell a story of amazing success. From 1922 to 1928 the index of industrial production climbed some 70 percent. Gross national product, measured in constant dollars, rose nearly 40 percent and per capita income about 30 percent."[9]

But all along there had been those weaknesses in the economy that we have seen: farmers, blacks, and a substantial number of ordinary workers were not sharing in the general prosperity. Per capita farm income was $223, as against $870 for nonfarm workers, and by 1929 some 6 million families were living on less than $1,000 a year.[10] Between 1919 and 1929 the output per worker in manufacturing industries had increased by 43 percent.[11] Wages and salaries had not gone up accordingly. Even by 1928 there was a glut of consumer goods coming into the nation's stores and shops. Manufacturers cut production and laid people off, which further reduced the country's capacity to buy. A downward spiral was set up, and as things turned sour a lot of banks, especially small ones in rural areas, collapsed. Everything that happened made everything else worse, and the juggernaut roared downhill until it could fall no further.

Behind it all lay something more basic: the prevailing belief that an unremitting attention to one's self-interest was an acceptable course to follow. Too many people saw the 1920s as an endless party. Speculators gambling with borrowed money with no concern for the consequences, industrialists holding the lid on wages, a government indifferent to the plight of farmers, and almost everybody unwilling to acknowledge the desperation in which most blacks lived—it was always me first, let the other man look out for himself.

This blindness was seen by some observers even at the time. John T. Gueenan, of President Hoover's Research Committee, writing in the early days of the Depression said, "Powerful individuals and groups have gone their own way without realizing the meaning of the old phrase, 'No man liveth unto himself.' " [12] The statement tasted a little of a jeremiad, but it was essentially correct, for it was not merely a few powerful people atop government and the corporations who were so absorbed in their own interests that they failed to see the cracks in the system, but the bulk of the middle class, whose influence far outweighed their numbers.

The human misery caused by the Great Depression was greater, I believe, than that caused by any other event in the country's history, with the exception of that suffered by some areas of the South as a result of the Civil War. By 1932 a quarter of American workers was unemployed. But matters were worse than this suggests, for three-quarters of those who had jobs were working part-time—either working shorter hours, or faced with chronic and repeated layoffs. [13] As a consequence over a half of the American population was living on sharply reduced incomes, and even those who continued to work as before could expect occasional pay cuts, loss of bonuses, and reduced commissions. As 1929 rolled into 1930 and then 1931, more and more stores stood empty along America's main streets. By the middle of 1932 industry was operating at less than half of the 1929 volume and wages were 60 percent less than in 1929. Business as a whole was running at a net loss of five billion dollars. [14] Farm income fell from $12 billion to $5.5 billion. [15] Frederick Lewis Allen says, "It marked millions of people—inwardly—for the rest of their lives. Not only because they or their friends lost jobs, saw their careers broken, had to change their whole way of living, were gnawed at by a lurking fear of worse things yet, and in all too many cases actually went hungry; but because what was happening to them seemed without rhyme or reason. Most of them had been brought up to feel that if you worked hard and well, and otherwise behaved yourself, you would be rewarded by good fortune. Here were failure and defeat and want visiting the energetic along with the feckless, the able along with the unable, the virtuous along with the irresponsible." [16]

The worst came in March 1933, at the very time a new President, Franklin D. Roosevelt, took office, when the American banking system collapsed, and all over the country banks were being closed by state governments. Now people could not get their hands on even what little money they had. But the arrival of a President with a "New Deal" raised hopes. Week after week, month after month, the new government shoved bold plans through a rubber-stamp Congress which was willing to give Roosevelt every chance to try his innovative schemes to save the suffering nation—mainly, of course, because few Congressmen had any better ideas themselves. The panaceas seemed, for the moment, to work. The banks reopened, times improved a little through 1935 and 1936; people began to talk of the Depression in the past tense. It was over—or soon would be—and the good old days would be back. The stock market began to march upwards.

Then in August 1937 the market broke, and by October was pouring downwards. It was 1929 all over again, if not quite on the same scale. The decline continued into 1938. The index of industrial production dropped from 117 in August 1937 to 76 in May 1938. In nine months two-thirds of the improvement gained in the New Deal years was lost.[17] Within a few months two million men, who had only recently gotten jobs, were once again out of work.

The recession of 1937–38 made it clear that there would be no easy end to hard times. And in fact there was not; the Depression continued into 1940, when the first effects of the European war were felt by the American economy; and it lingered on for another year or two until the wartime economy, coupled with the sopping up of the labor surplus by the draft, got industry working at capacity. The lesson was clear: nothing human beings could do had worked. Only with the impetus of a devastating war had the economy been revived.

The Great Depression of the 1930s was a disaster of appalling proportions. Perhaps half the working population at one time or another knew what it was like to lose a job. Millions actually went hungry, not once but again and again. Millions knew what it was like to eat bread and water for supper, sometimes for days at a stretch. A million people were drifting around the country begging, among them thousands of children, including numbers of girls disguised as boys. People lived in shanty towns on the fields at edges of cities or along railroad tracks, their homes built of packing cases, their food sometimes weeds plucked from the roadside.[18]

Given all of this, we might think that the country would have been struck by a revulsion against the excesses of the 1920s which were in good measure responsible for what had happened. We might expect that American people would turn away in disgust from the speculators who had gambled in people's futures, the big industrialists who had driven working people's wages down, the jazzers and speakeasy haunters who had flaunted their contempt for the older virtues of work and sobriety, the feminists who had abandoned domestic duties in order to smoke, drink, and fall in and out of bed as they liked, the writers and painters and critics who had sneered at the Victorian desire for an orderly society as hypocritical. In a word, we might expect a rejection of the cult of the self, the ideology of self-indulgence which had come to maturity in the 1920s.

But in fact it did not happen. What Americans wanted, despite everything, was more of the same. Let us look at some figures. We might begin with sex. Again, Kinsey's figures for the males are not broken down into cohorts by decade, but the figures for women allow us to discriminate between a "1920s generation" and a "1930s generation," that is to say, between those women who were coming into sexual maturity in the 1930s as against those who had arrived at that point a decade earlier.

We do not see remarkable changes. What we see, however, is a continuation of the movement which had begun in about 1915. The young women of the 1930s were beginning to pet a year or two earlier than their older sisters had, and at every age larger numbers were doing so. The same pattern was true of petting to orgasm: more of the younger group were doing it, and they were doing it earlier.[19]

Again, the change in the rate of premarital intercourse was slight, but it was nonetheless upwards. The taboo against sex before marriage for women was continuing to erode.[20] Inside marriage, more women were having coital orgasms, and the number of women who never reached a climax dropped dramatically in the younger cohort,[21] suggesting a more relaxed attitude toward sex. And finally, rates for adultery among women continued to rise, although only by a small percentage.[22]

The Great Depression, then, brought no reaction to what some had seen as the moral laxness of the Jazz Age Flappers. The effect was noted in Middletown. One professional observer told the investigators, "They've been getting more and more knowing and bold. The fellows regard necking as a taken-for-granted part of a date. We fellows used occasionally to get slapped for doing things, but the girls don't do that anymore." Another person commented, "Our high-school students of both sexes are increasingly sophisticated. They know everything and do everything—openly. And they aren't ashamed to talk about it."[23] Another of the townspeople said that by the early thirties, "There was a lot of sleeping about by married people and a number of divorces resulted."[24] Yet one more said, "Moral conditions are now generally at a low ebb. There is no disgrace attached to the things that cause divorce."[25] The Lynds go on to say that the collapse of sexual morality is hardly universal, but that it "does represent, however, a noticeable change in the total pattern of behavior,"[26] a conclusion that is supported by the Kinsey data obtained a decade later.

It is not surprising, then, that American divorce rates rose during the 1930s. The rate increased from 1.6 per thousand of population at the beginning of the 1930s to 2.0 at the end of the decade, a 25 percent increase.[27] In part the rise in the divorce rate can be attributed to the economic misery the country lay in—money problems are frequently a factor in divorce. But the crisis also produced countervailing pressures: many women, especially those with children, knew that they could not possibly support themselves and stayed in marriages they might otherwise have left; and some males, hard-pressed to find five dollars to feed their families with, could not possibly dig up the fifty dollars an uncontested divorce might cost.[28] The two factors probably came close to cancelling each other out; and still divorce rates rose.

So did rates of drinking. In 1934, the first complete year in which alcohol was sold legally, the rate was .97 gallons of absolute alcohol per capita.[29] However, this figure is for legal alcohol only, and is undoubtedly low. For one thing, the legal alcohol industry was in the process of getting on its feet; for another, it was necessary to develop a new sort of public drinking institution to replace the discredited saloon and the speakeasy; it took a little while for the "bar" and cocktail "lounge" to establish themselves.

In any case, by the late 1930s per capita consumption of absolute alcohol was back to 1.54.[30] This was substantially lower than it had been in the immediate pre-Prohibition years, but it was certainly higher than the level for the 1920s. And once again this was in the face of the economic crisis. With a fifth of the population out of work, and a considerable proportion of those who were working on short rations, discretionary spending was sharply curtailed, and non-existent in many families: drinking had become, for perhaps the majority of Americans, a luxury.

Yet consumption of liquor rose. One Middletown man said, "Drinking increased markedly here in 1927 and 1928, and by 1930 was heavy and open." The Lynds say, "On every hand, the testimony in 1935 was that there is much more drinking here now than ten years ago."[31]

Again, cigarette consumption continued to rise all through the Depression, and by the time of World War II was about twice what it had been at the end of the 1920s.[32] The Depression itself may have been a factor: a pack of cigarettes cost about a dime, and as a consequence smoking was one indulgence most people could afford.

In sum, the Depression brought with it no reaction to the new hedonism of the 1920s. Everything that was being done in the Jazz Age was being done in the 1930s a little more, and a little more frequently. People were drinking a little more, having illicit sex a little more, smoking a little more. They were, incredibly, still determined to make their fortunes gambling on the stock market, as the run up of stock prices in 1935 and 1936, and the subsequent crash indicate. And they were, as we shall shortly see, falling deeper in love with the entertainment industry.

CHAPTER 13

The Surrender to the Media

The importance of the rise of radio to the American culture was not so much for the pull it exerted on the American people, but for the fact that it set the form for what is probably the most significant innovation of the twentieth century, television. Nothing, perhaps, since the discovery of alcohol has so dramatically altered the nature of human consciousness as television. It has become, especially for Americans, a central—and for many people *the* central—element in their lives. For many Americans, television is what whiskey is to the social drinker—something they enjoy, but can do without. For others, however, it is drink to the alcoholic—something that lies at the heart of their days. And the beginning was radio.

The idea that sound could be sent long distances through the air without wires dates well back into the nineteenth century, and by 1900 it had been shown to be feasible.[1] Through the latter decades of the nineteenth century and beginning years of the twentieth century, technical improvements brought the broadcasting of sound to reality. By 1905 Guglielmo Marconi, who had developed a good many of the technical advances, had installed wireless radios on hundreds of ships, and many others had been established elsewhere.[2]

The new wireless system was given a great deal of publicity as the newest technical marvel of an age which had produced a good many of them, and inevitably all around the world amateurs were building for themselves wireless sets that could receive the sounds the air was rapidly filling up with. These sets were relatively simple to build from inexpensive parts which could be bought easily. Many of these amateurs went on to make sets that could transmit as well as receive, and very quickly there was created a network of "hams" who could not only

168

pick up the sounds in the air but add sounds to the increasing jumble out there.

It soon became clear that these thousands of radio transmitters would have to be organized to prevent utter chaos, and in that wonderful year of 1912 the United States government passed the first Radio Act to sort out some of the confusion. It remained the basic law of radio until 1927.[3]

Just as nobody had immediately seen that the primary function of the phonograph record would be to bring dance music into the homes, so it had occurred to nobody that the principal use for radio would be the dissemination of entertainment. It was to be used mainly, as befitted the Victorian mind which devised it, for serious, practical purposes by seamen, the military, governments and business, providing a much more portable and flexible means of communication than then existed. Paradoxically, for many of these purposes the principal problem was not how to reach the widest audience but how to beam to a specific target to ensure confidentiality. What was wanted was a wireless telephone. But once the problems were worked out, it seemed to many people, radio would make a great deal of money as a communications system parallel to the telephone.

Interest in the new device was given a boost by World War I, when thousands of American soldiers and sailors were trained in wireless radio. Many of them built their own sets when they came home, and then made others for friends who wanted to get in on the fad. By the end of the war complete radio receivers were available in stores across the United States. It is important to the outcome that receivers were in far greater use than transmitters. The great challenge for most early radio fans was to see how distant a signal they could bring in. It was a thrill to suddenly hear, faintly, the sound of foreign voices.

Then in 1919, when it appeared that British interests might be about to gain control of a good deal of the American market for radio, the United States government stepped in to act as midwife to the formation of a large radio combine.[4] This allied A.T.&T., which had bought up a lot of radio patents a few years earlier, and General Electric, to form the Radio Corporation of America. Radio was suddenly big business.

It was, however, still a business primarily of serious communication. The idea that it might become an electric version of vaudeville had not yet occurred to anyone. But the thought was not long in coming. By 1920 a number of hams who owned transmitters had taken to offering, simply for their own amusement, brief programs of music, usually provided by phonograph records. One early disc jockey was Frank Conrad, an employee of the Westinghouse Company in Pittsburgh.[5]

A Westinghouse executive named Harry P. Davis, whose role in the development of modern America would prove to be, by inadvertence, enormous, knew about Conrad's broadcasts. He noticed stories about them in the local press, and that a department store selling wireless equipment was advertising that they could bring in Conrad's "wireless concerts."[6] This department store advertisement, Davis later wrote, "caused the thought to come to me that the efforts that were then being made to develop radio-telephony as a confidential means of communication were wrong, and that instead its field was really one of wide publicity [i.e. wide dissemination]; in fact, the only means of collective communication ever devised."[7]

Davis's interest, of course, was in promoting the sales of radio equipment, not in producing a great entertainment media. It is likely that the same thought had occurred to others, and in any case would have sooner or later. But Davis was the first to act on the idea, and from that tiny, almost casual notion, grew the enormous broadcasting industry with its adjunctive cable and VCR systems which has come to dominate American life. When the history of modern times is written, the name of Harry P. Davis will have to be seen as being as critical as those of the Wright brothers, Edison, or Marconi. It is one thing to invent something; it is quite another to know what to do with it. We remember how scornful Edison was of using the phonograph record for preserving dialect jokes and bugle calls, and how he lost out to people who saw where the money lay.

Davis reasoned that radio concerts would attract a lot of publicity for the new device, and would at the same time give people some reason for buying one. He established Conrad in a building at the Westinghouse plant with a larger transmitter, with the plan of producing a regularly scheduled hour or two of evening radio programs, mainly concerts.[8] There was no thought that the concerts would earn money directly; only hopes of raising interest in radio.

Davis, who had a good mind for publicity, decided to open the radio station, which had the call letters KDKA, with reports of the results of the impending Harding-Cox presidential election in November. With the resulting blizzard of publicity, KDKA went on to build an expanded radio service. (Actually, as Erik Barnouw has pointed out in his standard history of broadcasting, the election results were also carried by 8MK, a Detroit station associated with a Detroit newspaper;[9] but KDKA got the publicity and the credit.)

Following the hoopla over the KDKA broadcasts, first scores and then hundreds of stations sprang up all over the country. By May 1922 the Commerce Department had granted more than 300 radio broadcasting licenses, and by the end of the year the figure had almost doubled.[10] In that year 100,000 receiving sets were sold; the next year it was 150,000, and more after that.[11]

The first radio stations were still not true commercial stations. A substantial number of them was associated with churches and universities and was offering programs as a public service. Most were established by newspapers, hotels, and department stores as publicity devices.[12] Shows ran heavily to music, sometimes from recordings but often live, featuring local talent—amateur pianists, singers, bands, and vocal groups. There were also "talks"—precursors of the present-day talk shows— and some of the more serious stations, especially those affiliated with the universities, offered dramatic readings.[13]

Broadcasting of entertainment was, however, still seen as a sideline, a diversion from radio's main task of long-distance communication for the use of business, the military, and the like. A great many hams were scornful of the frivolous entertainment broadcasts. Indeed, in 1919, when Lee DeForest, one of the major developers of radio, put some music on the air, he was visited by a government inspector who threatened to take away his license if he did it again.[14] But in the year or two after KDKA first went on the air with the election results, it became abundantly clear that radio was a simply gorgeous medium for publicity. And a number of people started looking around for a way to capitalize on this fact.

A question remained: How was it to be paid for? It was obvious that newspapers and department stores could not be relied on to subsidize the new industry indefinitely. A number of ideas was proposed: the federal government should build a national broadcasting chain—as would happen in Europe, e.g. England's famous B.B.C.—municipalities should create public service stations for their citizens—as many eventually did, e.g. New York's WNYC—the universities ought to take control. David Sarnoff, eventually to be the powerful chief of N.B.C., presciently suggested that six "super power broadcast stations" be established to blanket the country; and not surprisingly, an A.T.&T. executive suggested that his company develop a national network modeled on the telephone system.[15]

However it was to be done, it was generally believed that the "air waves" were a public property and that the new means of communication ought to be run for the public good. The reasoning was that, unlike the case of newspapers, only a finite number of radio stations could be established; surely these should not be put into the hands of private interests to do with as they liked. And this was the official position of the United States government. In 1922 Herbert Hoover, who, as Secretary of Commerce, would play the largest role in shaping government policy toward the new device, said, "It is inconceivable that we should allow so great a possibility for service to be drowned in advertising chatter."[16] In retrospect this was possibly the funniest statement offered by a public official of the period, although there were certainly many contenders.

But Herbert Hoover had started life as a mining engineer and a businessman, and at heart he believed in laissez-faire capitalism. He said that government must "establish the public right over the ether roads";[17]

but in fact, he dithered. Among other things he permitted the sale of radio stations,[18] which effectively made them private property. In fact, there was no real government policy toward radio; it did nothing but distribute the "air lanes" in order to prevent chaos, and for the rest of it sat on its hands. There was a vacuum, and inevitably people who saw ways to make money from the medium rushed in to fill it. And a great opportunity was missed.

The real beginning of commercial broadcasting came in 1922, when A.T.&T. hooked together a number of stations by way of its telephone lines, with WEAF in New York as the centerpiece of the system, and began to sell advertising time. Hoover's Commerce Department "gave its blessing"[19] to the scheme, despite what Hoover had thought about drowning the air in advertising chatter. Radio was to become a sort of "telephone booth of the air,"[20] providing access to the public to anyone who could pay for it.

The stations, at first, did have some qualms about allowing makers of toothpaste and soap to introduce long sales pitches in the middle of programs. In particular, on the first commercial broadcast issued by the WEAF combine, on August 28, 1922, at five in the afternoon, a Mr. Blackwell delivered a long monologue on behalf of Hawthorne Court Condominiums in Jackson Heights in the New York City Borough of Queens. Station heads were appalled, recognizing that if they allowed anybody who had the money to pay for it to make long speeches on behalf of this or that, they would shortly have no audience.[21] They began limiting the length and content of commercial messages, with the result that advertisers, to increase their exposure, began naming their shows after their products. By 1927 the country was blessed with the Maxwell House Hour, the Ampico Hour, the Palmolive Hour, the General Motors Family Party, the Cities Service Orchestra, the Sieberling Singers, the Ipana Troubadors, and countless others.[22] At the same time station chiefs realized that in order to hang onto their audiences they had to offer "sustaining" shows during the times they had no commercial programs to broadcast. Through the decade the number of advertisers, and the length of advertisements, increased. And in less than ten years from the date of the first regular program, the "great possibility for service" had become almost entirely trivialized.

In the end, of course, the fault lay with the American public. Far from protesting what was happening, it lapped up the seemingly free flood of entertainment. Sales of radio equipment went from $60 million in 1922 to $136 million in 1923, to $340 million in 1925, to $650.5 million in 1928. By 1927 there were radios in cars.[23] Radio personalities were becoming celebrated, were being met, when they traveled, by bands and mayors giving out keys to the cities.[24] In 1926 the National Broadcasting Company was formed; the competing Columbia Broadcasting System was created not long after, and national radio was now a fact.

Radio receivers were becoming more sophisticated. They were being designed as pieces of furniture to fit into living room decors; improvements were made in sound quality; and in 1928 the receivers were adapted to house current. "By 1930 some six hundred radio stations were broadcasting to more than twelve million radio homes, about 40% of the total number of American families," according to one authority, Daniel J. Czitrom.[25]

But the real "Age of Radio" came with the Depression, and lasted for about twenty years through to the moment when television shouldered it into a secondary role in the American media system. The Depression, all in an instant, priced a lot of the old show business beyond the means of millions of Americans. A seventy-five-cent record was now a luxury and so was a visit to a vaudeville theater, or a dance hall. Vaudeville, which had been sick for some time, simply gave up and died. The phonograph industry collapsed completely: sales dropped from $104 million in 1927 to $6 million in 1932.[26] Scores of record companies went bankrupt, in the end leaving only Columbia and Victor (the cut-price label Decca was formed in 1934), and both were on the edge of bankruptcy. The night clubs died, and so did the dance halls. In November 1930, *Variety* carried the headline, "Dance Halls All Starving," and the piece went on to say that it was "the worst season they have experienced since the war."[27]

But radio programs, once you had bought a set, were free, so people sat home and listened, and very quickly the "radio habit" was formed. The Depression, then, was in part responsible for the sudden prominence of radio in American lives.

Not coincidentally, the nature of radio changed in the beginning years of the Depression. Until then, music had been the staple of the business: news, talk shows, drama, played secondary roles.[28] The most important function of radio was to provide concerts of music of various sorts, and especially music for dancing. Now, at the beginning of the Depression, it was discovered that audiences were eager for stories. The catalyst was a show featuring two whites in black-face—or black-voice, to be accurate—named Freeman Fisher Gosden and Charles J. Correll. These two men had put together an act which had been working fraternal lodges, carnivals, and the like in the waning days of vaudeville, playing two amiable "darkies" named Amos and Andy in stereotypical dialect replete with solecisms, some of which, like "I'se regusted," went on to become national catch phrases. The *Amos 'n' Andy* show began in the late 1920s and by the early 1930s was the most popular radio program in the world and possibly—in terms of relative population size—the most popular radio show ever. One survey claimed that half of

Americans regularly listened to *Amos 'n' Andy*. Some factories even
changed their working hours to accommodate *Amos 'n' Andy* fans among
their employees.[29]

As offensive as the concept of *Amos 'n' Andy* may seem today, it must
be admitted that it was extremely well done. The two leading figures,
along with a raft of subordinate characters like the Kingfish and the
wives of the proponents—all played by Gosden and Correll—were deftly
outlined, and the rambling saga, which ran on week after week, was
sufficiently compelling to keep its huge audience coming back. By the
end of the 1920s Gosden and Correll were earning $100,000 a year, an
enormous sum for the time. Erik Barnouw says,

> The *Amos 'n' Andy* series broke new ground in several ways. It established
> syndication as a mechanism even though recordings were still of doubtful
> quality and limited to five-minute lengths. The feasibility of the continued
> story was also overwhelmingly shown. A basic dilemma continued for weeks,
> far from alienating listeners, enmeshed ever-widening rings of addicts . . .
> All this would bring a flood of serials in succeeding years.[30]

A flood it certainly proved to be. Throughout the Depression decade
they marched in a mob through American homes: The Goldbergs, Little Orphan Annie, Skippy, Just Plain Bill, Myrt and Marge, Buck Rogers, Vic and Sade, Fu Manchu, Charlie Chan, the Shadow, and dozens
of others.[31] These dramas characterized radio in its so-called Golden
Age and laid the foundation for the basic material of television: soap
operas, situation comedies, and cops and robbers shows.

But the dramas had competition. Almost as popular, although fewer
in number, were the "comedy" shows, which were in fact small variety
bills employing the talents of the best of the vaudevillians, among them
Eddie Cantor, Ed Wynn, Ken Murray, Al Jolson, Fred Allen, Burns
and Allen, Stoopnagle and Budd, and, perhaps the most famous of all,
Jack Benny.[32] The format for these comedy shows was much the same
for all: a lead figure with some well-defined character trait, like Allen's
cynical acerbity, Benny's stinginess, Gracie Allen's daftness, along with
a group of subordinate characters whose foibles could be made a source
of banter. Each show usually included slices of verbal byplay, a skit, a
musical number or two, and often a guest star who did a brief turn.

As the decade progressed, sports played an increasingly larger role
in programming, in the main the famous play-by-play baseball and college football broadcasts. Music remained important: swing bands were
a staple of evening radio, and surprisingly, there remained a large audience for opera and symphony.[33]

And as the coming debacle in Europe began to rise over the horizon,
Americans turned more and more to radio to hear the news of the
great events rolling toward them. By 1939, according to one study, 70

percent of Americans "relied on radio as their prime source of news, and 58% thought it more accurate than that supplied by the press."[34]

But at the heart of radio were the dramas: by 1940 soap operas constituted 60 percent of daytime programming, and the average soap fan followed 6.6 different shows.[35] What radio listeners wanted most of all was fantasy: the invisible Shadow, the all-powerful Daddy Warbucks, the sleuthing genius Charlie Chan, the space adventurer Buck Rogers with his ray gun and spaceship. Even the domestic comedies and dramas, like Vic and Sade, Ma Perkins, and the soap operas, were essentially fantastic, about people doing and saying things that people wouldn't, or even couldn't, do.

The devotion of radio to fantasy was occasioned in part by the belief of the people who ran it that Americans did not want to be reminded of their troubles by long discussions of the problems of the unemployed and the desperate, but preferred escape. But there was another factor: by 1931 the business community was in control of radio and did not want to fill the air with a lot of comment on what seemed to be the collapse of capitalism. It was not simply that the advertisers had a veto over what went into the shows they were sponsoring; it had gone further than that. By this point most of the major shows were actually being developed and produced by advertising agencies based mainly in New York, but in a few cases in Detroit, Los Angeles, and elsewhere.[36] It was not just a question of ideas being censored; in the main nobody involved had the faintest notion of producing the kind of show in which dangerous ideas would come up. The radio stations more and more became the landlords of air time, with little say over what went out from their studios. The consequences of this system were enormous: automobile manufacturers, purveyors of soap flakes and breakfast cereals and gasoline were deciding what the American people heard on the new medium that occupied so much of their attention. One result was that the greatest news story of the day, the Depression, was virtually ignored. WLW in Cincinnati, for example, forbade references to strikes in its news shows.[37]

Thus, within a decade of Herbert Hoover's pronouncement that government must "establish the public right over the ether roads," the whole idea of radio as a public property had dried up and blown away in the wind. To be sure, lip service was paid to it in the notion that stations must provide a certain amount of public-service programming. But by and large the stations were left to define for themselves what constituted public-service programming and the result was that radio became what television would become: a system for bringing salespeople into American homes.

If the radio boom in America had come ten years earlier, when the Progressives were dominant, or ten years later, when Roosevelt and the New Dealers were in power, it is probable that things would have turned out differently. In both the Progressive Era and the New Deal years, there was substantial anti-business sentiment among the middle class, in academia, in the labor movement, and even among holders of high office. In either period a campaign to keep broadcasting out of the hands of business would have had popular support. But the radio mania came in the 1920s, when the stock market was running haywire, the government was strongly pro-business in philosophy; when a Bruce Barton would have a huge success with a book that presented Jesus Christ as a super businessman.[38] The businessman, millions of Americans believed, epitomized all that was right about America. Corporate presidents were being written up in magazines as oracles whose views on almost any subject had to be weighed heavily. In the 1920s there would be no campaign to keep business from gaining control of the broadcast medium.

There was then, and still is, a fairly large minority of Americans who objected to the vulgarity, commercialism, and other offenses of American broadcasting. It has often been a vociferous and vigorous minority, and it has sometimes had some effect in curbing the worst of the offenses. But it has been, nonetheless, a minority. The simple truth is that, by and large, the public has liked what broadcasting has been giving them.

The Depression, then, did not cause the people to rise to their feet shouting for government to loosen the grip of business on radio; the people called, instead, for more of the same. And they did exactly that as well in the case of another element of the still expanding entertainment business, song and dance. There was no turning of backs on the jazz music and the sexy dances that were done to it which symbolized the self-indulgent age which appeared to have brought on the Depression; it was more of the same.

The music of the Depression period was somewhat different from that of the 1920s, and so were the popular dances, but essentially they were modifications of, or if you wish improvements on, what had gone before. With the collapse of the economy, followed by the general bankruptcy of the recording industry, and the shuttering of the dance halls and the cabarets, it appeared initially that the jazz mania of the Roaring Twenties had cured itself. The people struggling to keep the music industry alive concluded that what people wanted was no longer the hot stuff of the jazz bands, but dreamy, sentimental music about small hotels by waterfalls, and love that went on forever.[39] This view was partly correct—but then, there has always been an audience for sentimental music, as the long and successful career of the Guy Lombardo Orchestra makes abundantly clear. And it is also true that the hot, small dixieland band, out of which jazz grew, had fallen out of favor, except among a small handful of jazz fans.[40]

But the taste for hot music had not, especially among the young, died out. A surprising amount of excellent jazz was recorded during the early days of the Depression, perhaps simply out of momentum— among them a number of masterpieces by Louis Armstrong, which were selling at the rate of 100,000 a year,[41] an exceedingly good figure for Depression times.

Then in 1934 the National Biscuit Company, in order to promote a new cracker called Ritz, instituted an ambitious three-hour show called "Let's Dance," which was to feature a sweet band, a Latin band, and a hot band. Almost by chance a new band which had recently been put together by an admired free-lance clarinetist named Benny Goodman was chosen as the hot band. Between the exposure the show gave him, and growing record sales, the band had by the middle of 1935 developed a following on the West Coast. Its stay at the Palomar Ballroom in Los Angeles was a roaring success,[42] and thereafter the Goodman orchestra rose swiftly into popularity, to a moment in the spring of 1937 when, during an engagement at the Paramount Theater in New York, young people began jitterbugging in the aisles, with much attendant publicity.[43] Goodman was a celebrity and growing rich; and on the heels of his success dozens and then scores of other bands were formed in hopes of catching the same lightning.

Swing music of the 1930s was an outgrowth of the music created by the big jazz-oriented dance bands of the late 1920s. Especially influential were the orchestras of Ben Pollack, Fletcher Henderson, Jean Goldkette, Red Nichols, the Casa Loma, and McKinney's Cotton Pickers. The music of these bands had, in turn, been created by combining the jazz feeling and improvised solos of the small dixieland bands with the structure of the big dance bands, which interspersed solos with arranged passages.[44] The leading example was the Paul Whiteman Orchestra, but there had been others. The swing bands essayed a somewhat lighter approach to rhythm, a smoother, more flowing style, and many "riffs" or short repeated figures to develop rhythmic momentum. But the basic nature of the music was the same, although dance band fans could tell the difference between the older and newer styles.

After Goodman's success in 1935, and particularly after the publicity attendant upon the uproar of the 1937 Paramount Theater engagement, swing became a national fad, with band leaders like Tommy and Jimmy Dorsey, Glenn Miller, Duke Ellington, Artie Shaw, Woody Herman, and Count Basie becoming celebrities and earning substantial fortunes.[45]

The life of the swing fad was brief, ending in about 1945. But while it lasted, swing dominated popular music. Developing along with it were styles of dress (the bobby sox and eventually zoot suits), an arcane lingo (fans were hep-cats, hot riff tunes were killer-dillers), and dances (the Suzie-Q, the Big Apple).

It is important for us to see how similar the youth cultures of the

178 / The Rise of Selfishness in America

Jazz Age, the Swing Era, and the Age of Aquarius of the 1960s were beneath the superficial differences. All three had their theme musics, the associated dances, the styles of dress, the argot. All three appeared on the surface to be ages of rebellion, but were in fact building on ideas, themes, modes, and attitudes developed by earlier generations, as we shall see more clearly when we come to deal with the 1960s. And underlying all three was a hedonism justified by a philosophy of "freedom" which demanded as a right that people should "be themselves," "express themselves," or, as the 1960s put it, "do their thing."

The primary difference between the Swing Era and the others was that the 1935–45 period was overshadowed by the Depression and World War II. When a quarter of a nation was unemployed, or primarily occupied with fighting the most extensive and bloodiest war mankind had ever seen, cries for freedom and expressiveness—the demands of the self—had to be muted. But they were there nonetheless.

So was the desire to be entertained. Despite the phenomenal rise in popularity of radio, the movie-going habit which had been created in the 1920s continued into the 1930s unabated. Film had been given an enormous impetus in 1927 with the arrival of sound. From the early days of the movies' popularity people had been searching for a way to join recorded sound to film. Most of these efforts had turned on the idea of synchronizing something like a phonograph record to a film, but this had proven difficult to work out, especially as phonograph needles were prone occasionally to skip grooves and film to break.[46] The answer lay in trying to find a way to attach the sound directly to the film, although for a while sound on disk continued to have some support. In time Lee DeForest and others found ways of doing this, and in 1926 Warner Brothers produced three feature films that included music. Then, in 1927, "a sentimental, turgid drama"[47] featuring Al Jolson, called *The Jazz Singer,* which also included music, was a huge commercial success. From that point on sound rapidly killed off silent film: by 1929 some 9,000 movie houses had converted to sound.[48]

At the same time other technical advances in lighting, more moveable cameras, new editing and cutting methods helped to make films more flexible. Particularly important was the development of feasible means of reproducing color, which gradually came to supersede black and white step-by-step through the 1930s and 1940s.[49] These technical improvements, too, helped to keep audiences coming back. Now the movies were really "real." Initially attendance dropped as the Depression got under way; but by 1940 it was almost 50 percent higher than it had been in 1932.

The movies of the period were dominated, as has often been the case, by the so-called "genre" films—movies made to fit into specific categories, each with its own conventions: musicals, horror, western, and the like. Because these movies were cut to a common pattern, often even using not merely similar but the same costumes, backgrounds, and actors, they could be quickly and cheaply made. John L. Fell says: "If audience expectations followed patterns established by convention, then attendances might be assured when the expectations were satisfied."[50] Not every genre film was so stereotypical; within each category there occasionally appeared films which bear watching today, in the view of film critics: *The Virginian,* among westerns, *The Island of Lost Souls,* in the horror category, the *Thin Man* series among detective films.[51]

But for the most part the movies, like the radio dramas, were fundamentally fantasy. They were made explicitly and deliberately to refer to other movies, not to real life. The detective movie almost always ended with the sleuth in a room full of suspects—frequently at an upper-class dinner party—discovering the murderer by deduction. The musical again and again told the story of the unknown singer or dancer stepping into the lead on opening night when the star breaks his or her leg. Conventions applied to tiny details: at the requisite final dinner party, the murderer "is always to be seen behind Nick Charles . . . usually behind his left shoulder. . . ."[52]

Thus, although some fine films made during the period were box office successes, the largest part of the public's diet consisted of genre quickies featuring celebrated stars playing out the conventions of the type. The function of the movies, as was the case with "Ma Perkins" or "The Shadow" on radio, was to remove the viewer into a fuzzy world of make-believe inhabited by people doing things people not only ordinarily did not do, but in most cases could not do.

The appetite for fantasy exhibited by Americans in the 1930s in the hours spent at the movies, listening to their radios, and in a good deal of their reading, was far greater than it had been heretofore.

It is hardly surprising that the American people, faced with unemployment or the threat of it, dwindling incomes, and perhaps more significant for many, the loss of hope, of faith in the future, plunged head first into the fantasy world offered by the new entertainment media. Who could blame a young couple unable to marry because they could not afford a place in which to live together for wanting to watch Fred Astaire and Ginger Rodgers dance in luxurious penthouses, or on the decks of opulent ocean liners? Who would deny the unemployed working man, dependent on the charity of his in-laws for the support of his children, his moment of solace with *Amos 'n' Andy?*

Yet there is something a little strange in the spectacle of so many Americans spending so much of their time and money on amusing themselves when the world was going to pieces around them. During the desperate days of the Depression one of the few industries that not only survived but prospered was show business. While the doors of factories, offices, shops and stores were closing almost by the minute, the movies, radio, the music trade, and by the later years of the Depression even the record business, were making huge fortunes for those lucky enough to have found places in them. Actors, movie producers, band bookers, radio announcers, script writers were buying country estates, Cadillacs, race horses, as once-comfortable accountants, real estate brokers, and apothecaries put their houses up for sale on a dead market and brought their children home from college.

It is true, of course, that the Depression created a groundswell for, if not socialism, at least social reform; and it is also true that millions of people became deeply involved in unionism, political action, groups and causes aimed at producing a more equal distribution of wealth and a general improvement in the lives of working people. There was a measure of idealism in the air.

Nonetheless, despite the willingness of millions of Americans to spend their spare time at union meetings, in political halls, and on street corners handing out leaflets, it is a matter of statistics that most Americans were devoting increasing proportions of their time to satisfying the needs of the self. For most, the response to the cataclysm of the 1930s was not to march, but to go to the movies. The new ethic of the self as a primary concern was now firmly established; and if the stock market crash and the economic disaster that followed could not dislodge it, what was likely to?

CHAPTER 14

The Incredible
Post-war Prosperity

The story of the cataclysmic war of 1939–45 has been told again and again. Indeed, the literature on World War II is so enormous that it would be a lifetime's work to absorb it all. Books on the subject were best-sellers when the war was hardly begun: *See Here, Private Hargrove*,[1] a lighthearted view of training camp life, was one of the biggest hits of the time, and many accounts of battles, like *Guadalcanal Diary*,[2] and *They Were Expendable*,[3] were best-sellers before the war was over. Books on the war have continued to command large audiences ever since: Cornelius Ryan's *The Longest Day*,[4] about the Normandy landings, William L. Shirer's *The Rise and Fall of the Third Reich*,[5] and many books on the surprise attack at Pearl Harbor have appeared without a let up.

There have been, however, a number of shorter histories of the war; readers who would like more information on the subject than I can give here might look at one of the more recent ones, James L. Stokesbury's *A Short History of World War II*,[6] on which I am depending for much of my brief sketch of the fighting.

The war was set off on September 1, 1939, when Hitler marched into Poland. The English and French, who had again and again accepted Hitler's annexation of territories belonging to other nations were finally forced to face up to reality, and on September 3rd declared war.[7]

Watching from across the Atlantic, Americans were not quite sure what to do. There were plenty of American men still in their forties who had been mobilized two decades earlier to fight the Germans, and furthermore, although the general public had only a vague awareness of Adolf Hitler's evils, there was a sense that he was not a very nice man. As a consequence, public opinion favored the British and what remained of their allies.

But there was a strongly pacifist mood in America at the moment. Contrary to what is widely believed, Americans are not a naturally violent people always thirsting for combat. Taken over all, since the founding of the nation, the United States has one of the best records on war and peace of any major industrial nation. There has only been one major conflict on the North American continent since the American Revolution—the North-South conflict—and the nation has lived in peace with its bordering nations for something like 150 years. The quarreling among Europeans occasioned no sudden uprush of desire to join in. A good deal of publicity had been given in the 1930s to the huge profits made by the armament manufacturers during World War I, who were being seen by many Americans as "merchants of death."[8] Furthermore, books and plays like Eric Maria Remarque's *All Quiet on the Western Front,* Ernest Hemingway's *A Farewell to Arms,* and John Dos Passos's *Three Soldiers*[9] had torn the romance from battle, showing it in all its gory horror. Nobody wanted a repeat of that. Indeed, a Gallup poll at the time indicated that 70 percent of Americans thought it had been a mistake for the United States to have entered the previous war.[10]

Whether the United States would have been drawn into the European war is hard to know, but in the event the country was given no choice in the matter. Through the 1930s the Japanese, faced with a large and growing population confined to a chain of small islands, and lacking many of the raw materials on which modern industrial societies are built, had been looking convetously at the vast, mineral-rich lands around her.

The Japanese did not think they could conquer the United States. Their intention was simply to drive the westerners out of the area in question. They assumed that they could do this with a quick strike, and that when faced with a *fait accompli* the unprepared and indulgent Americans would accept the situation rather than fight back.[11] And so, on December 7th, 1941, the famous day of infamy, the Japanese bombed the American navy at Pearl Harbor, and then leapt off to drive troops into Singapore, the Philippines, and the areas around them. Within six months they had gained most of what they had wanted. The United States was now at war, willy-nilly, not only with Japan, but with the other "Axis" powers, Germany and Italy, who also declared war on America.[12]

What happened next was one of the most astonishing performances of the industrial age. The United States began to mobilize, turning civilians, most of whom had never fired a gun in their lives, into soldiers and sailors almost overnight. Even more astonishing was the way the American industrial machine, which had been working at half-speed for a decade, was suddenly brought up to full throttle. Not only did the United States have to build itself a whole great war machine, but it had to keep its allies, at this point Russia and England, supplied with airplanes, guns, tanks and everything else necessary to fight the war. Yet in less than a year after Pearl Harbor, the United States built a sufficiently large and competent fighting force to be able to join England in an invasion of North Africa, which in time drove out the Germans.[13]

From this point on, although there would be difficult moments, and a lot of hard, dirty, inglorious fighting, it was mostly all victories. By 1943 it was clear that the tide had turned, with American forces driving forward island by island across the Pacific towards the Japanese homeland, and the Allies pushing—albeit slowly—up the Italian boot, and driving the Germans out of Russia. On June 6th, 1944, the Allies landed on the Normandy beaches to begin the longest day, and by the end of the year it was clear that the Germans were finished, although there would be one last strike out of the death throes with the attack through the Ardennes in what is now known as the Battle of the Bulge.[14] In May 1945, the war in Europe was over.

The fight in the Pacific went on, and it appeared that it was going to take some extremely bloody work to dig the Japanese out of the island stepping stones to their home islands. But then on August 6th, the United States exploded an atomic bomb over Hiroshima, and very quickly thereafter the Japanese surrendered.[15]

Although not immediately apparent, the war brought about a massive shift in the political structure of the world. European dominance of large portions of Asia and Africa was ended, although the French, English, and Dutch for a while struggled to hang on to their old colonies. The center of power had moved away from Europe to the United States and Russia, and the formerly great world powers, especially England and France, had to deal with the unpleasant fact that they were now perceived as secondary nations. The "American Century" it appeared, had arrived.

In the minds of most people, including many historians, the "post-war era" in the United States is seen as consisting of two antithetical periods. The first was "the Fifties," a time of homes and families when mothers stayed home and cooked and looked after the kids; a time when everybody was buying ranch houses in thousands of "Levittowns" across the United States; a time when teenagers built hot rods and drank Pepsi instead of beer, smoked Lucky Strikes instead of crack. The second period was, of course, "the Sixties," which can be dated roughly from the assassination of President Kennedy in 1963 to the sudden political and economic changes brought about by Watergate, the growing unpopularity of the Vietnam war, and the end of post-war prosperity due to a number of complex forces.

To the contrary, I hope to show that the cultural process which took place in the second half of the twentieth century in America paralleled, to an astonishing degree, a similar process that occurred in the first half of the century. Just as the ideas of the 1920s had been preceded by earlier attitudinal shifts, so we can find in the 1950s all the ideas that seemed so revolutionary in the so-called Sixties. And, further, just as the shift in behavior presaged by the ideas of 1912 and thereabouts did not spread through the culture until the late 1920s, so the revolution in behavior which is thought to have occurred in the sixties in fact did not take place until the 1970s. It does seem clear—in these cases, anyway—that widespread behavioral changes do not take firm hold until five to ten years after the necessary changes of attitude.

In this view, the post-war period divides itself sharply in about 1973. From the end of World War II until that moment there was considerable continuity in America in politics, economics, and "lifestyle," or behavior. Politically, liberalism dominated with a succession of Presidents from the liberal wing of the Democratic party—Truman, Kennedy, and Johnson. Even the Eisenhower presidency was fairly liberal in practice. Economically there was a gradual, but steady, rise in prosperity, with inflation fairly well controlled and employment high. Behaviorally, as we shall see in some detail, there was a continuation of the broad, slow movement, which had begun decades earlier, towards increasing permissiveness in sexuality, wider acceptance of the consumption of alcohol as a human norm, greater involvement with the entertainment media, and in general, a continued insistence on the primacy of the self. By contrast, in the early 1970s there was a swing politically to the right, the ending of post-war prosperity, and a dramatic upward surge in selfishness which very quickly became so gross as to effect a qualitative change in the nature of American life.

The most significant fact about the first of these two post-war periods was its astonishing prosperity. During the war the United States government, faced with the need to create a huge arsenal almost overnight, was prepared to spend anything necessary to get the job done. And it did: federal budgets, which had been running around $7 billion to $9 billion a year during the Depression, leapt to $98 billion by 1945. The national debt soared from $40 billion to $269 billion in 1946.[16] The government was pouring money through the system by the bucketful, and it fell like a shower of gold onto a people who only a few years earlier had been scrambling desperately just to pay the rent.

Happily, the chief beneficiaries were those who deserved—and needed—it the most. High taxes kept the rich and the manufacturers who controlled the industrial plant from skimming off the cream, as they had so often in the past. Salaries of the middle class were held in check by wage and price controls.[17] The real gainers were the farmers, technicians, and skilled laborers. Farmers certainly deserved a break. They had suffered from low prices through the 1920s, then even lower prices and successive years of drought during the 1930s. Now they could sell anything they could grow at good prices, and by good luck the string of droughts ended. Farmers began to pay off mortgages and put money in the bank.[18]

Similarly, anyone with a technical skill was, in the war years, in great demand. Machinists, draftsmen, carpenters, glassworkers, electricians—these people, once merely hired hands to be employed as cheaply as possible, were, as the industrial machine began to run at high speed, kings of the labor force.[19] In theory their wages were controlled, but in fact ways were found around the controls by employers desperate for their services. With bonuses for overtime, people who in 1940 were struggling to make $50 a week sometimes found themselves drawing paychecks for ten times that amount. Between 1939 and 1945, pay in manufacturing rose by 86 percent against a 29 percent rise in the cost of living.[20] Furthermore, these skilled people were far too valuable in the war plants to be sent off to carry M1 rifles in the swamps of New Guinea. And after so many years of famine, for four years they feasted.

The consequence was that the affluence created in the United States by World War II reached all the way down to the bottom of the economy. To be sure, even with high taxes sucking up excess profits, owners of even small manufacturing concerns, frequently operating under government contracts which guaranteed them a profit, could not be kept from amassing fortunes. But for the first time in the history of the American industrial society nearly everybody got a piece of the pie. Even the blacks, notoriously last hired and first fired, were getting a share.

George Katona, who produced a series of analyses of the American consumer during the post-war period, says, "Before 1940 American consumers as a whole had insignificant amounts at their disposal beyond what was calculated to be needed for basic living costs."[21] In the 1940s, through the war and the immediate post-war years, the money available for discretionary spending went from $40 billion to $100 billion, according to Katona, and by 1959 it was $200 billion.[22] Or to put it the other way around, the numbers of Americans with discretionary income to spend rose from 12 percent in 1929 to 25 percent at the end of World War II.[23]

The important thing about this sudden wartime prosperity was that there was little to spend it on. No new cars were being made, houses were not being built in large quantities, major appliances were generally not available. There was not much to do with spare money except to save it, and there was constant pressure on people to buy war bonds. Once it became clear that the war would be won in a few years, people began putting money aside to buy the new cars, new refrigerators, new washing machines, new radios, they expected would become available once the war was over. The country came out of the war with billions of dollars in savings. In 1939 only 10 percent of American families had bank deposits or government bonds; in 1946 the majority of families had one or both.[24]

When the war ended, and consumer goods came on the market again, people began to spend. The country thus averted what everybody had feared, that the cessation of wartime spending by the government would produce a sudden depression. Consumer spending took up the slack, and the economy remained buoyant, although with considerable inflation and momentary ups and downs.[25]

Then, in June 1950, the Communist armies of North Korea suddenly plunged southward. South Korea's army collapsed, and the North Koreans rushed down the peninsula toward Seoul, and then beyond. A hastily thrown-together force managed to stop the North Koreans at the bottom of the peninsula. Once again America mobilized, and along with troops from a number of other countries eventually managed to stabilize the battle line around the 38th Parallel.[26]

Once again a relatively few Americans paid the price for pushing the country's prosperity to yet new heights. Says historian Eric Goldman, "During the Korean War the whole United States was one big boom, the boomingest America in all the prosperous years since V-J. Virtually all of the soft spots that were still left in the economy were removed by the Korean War. The most telltale sign, employment, told a tale of historic proportions. Two months after the Korean War began employment crossed the sixty-two million mark—two million beyond the fondest dreams of the New Dealers at V-J. As a matter of fact, by August 1950, New York State had so few unemployment claims that it fired five hundred people from its compensation division."[27]

The people who were enjoying the prosperity of the war and immediately post-war years remembered the Depression with considerable clarity. Many of them had been harshly bitten by the experience in youth and early adulthood. In 1945, when the war ended, every adult American had seen the Depression at first hand, and most had been hurt by it one way or another. The majority remembered the stock market crash. And all of them had had their lives interrupted, and in many cases substantially damaged, by the enormous war that had just ended. They had seen fifteen years of cataclysm and catastrophe, and it is hardly surprising that what they wanted was calmness and order. The older of them wanted to pull together careers, or find secure jobs with some sort of pension visible down the road; the younger ones wanted to get into the job market, marry, raise families, if possible in more comfort and prosperity than they had grown up in themselves.

These sounded like modest enough goals, but at the time none of these post-war adults were at all sure that they would reach them. They had seen it all go sour before, and they were nervous that it would happen again. Everything they had experienced had taught them to be cautious, conservative, to value things like a decent home, a good marriage, well-behaved kids. And this was what they set out to get.

To this extent, then, the popular view of the Fifties as a time of homes and families is true. People were getting married younger; the median age for marriage for men went down from 24.6 in 1920, to 22.7; for women it dropped from 21.2 to 20.2. Once married they were having children quicker, too: in the early 1930s about a quarter of couples had children in the first year of marriage; in the years right after the war it was a third, and by the second half of the 1950s it was 44 percent.[28] They were having second children quickly, too: within the first three years of marriage during the Depression wives were producing about 0.75 of a child each; in the 1950s it was more than one each.[29]

This rapid family formation created a demand for new housing, and the construction industry responded. The first suburban boom had come in the 1920s, when Grosse Pointe Park grew by 724 percent, Shaker Heights by 1,000 percent, Beverly Hills by 2,485 percent. "The decade from 1920 to 1930 saw the complete emergence of the modern residential suburb," says one observer.[30] The move to the suburbs was slowed by the Depression and halted by the war; until the post-war period suburban living was available only to the relatively affluent. But on the tide of postwar prosperity a flood of people was swept out of the inner cities. Young people who had grown up in cold water flats in New York's Little Italy, Chicago's South Side, hungered to see their children passing footballs on well-mowed lawns, walking to good schools along quiet, shady streets. They were enabled to do so by the development of standardized housing, made from interchangeable parts constructed in factories and trucked to the building sites. Say Steven Mintz and Susan Kellogg, "Up until World War II the best educated and most prosperous Americans tended to live in cities. . . By 1960 as many Americans lived in suburbs as in central cities."[31] In the decade of the 1950s the population of the suburbs increased by 50 percent, that of the inner cities by only 11 percent.[32]

In part this rush out to the suburbs was driven by a broad-based movement of people whose parents had been blue collar into the white-collar class. In assessing a movement like this a great deal depends on how each of the classes is defined. But according to one authority, 18 percent of Americans were white collar in 1900, 31 percent in 1940, 37 percent in 1950, and 42 percent in 1960.[33] This growing middle class was made possible by changes in the industrial structure—the shift from manufacturing to services, increasing automation which required fewer workers on the shop floor and more in front of typewriters and the flow charts, and other factors. To some extent the change was more visible than actual: many of the people who now wore suits to work

188 /   The Rise of Selfishness in America

instead of overalls were doing work that was not much less routine. But putting on the suit and the tie automatically conferred both a certain status and a higher wage than overalls did.

A second factor in this burgeoning of the middle class was the so-called G.I. Bill of Rights, under which the government subsidized college educations for almost anyone who wanted one and was capable of doing the work. College enrollments soared.[34] People were being trained to fill the executive and professional jobs that were being created.

But in the years immediately following the war, the shadow of the Depression still fell across post-war prosperity. Most people remained wary. In the early 1950s only 10 percent of Americans owned any stocks; their money was in savings accounts and government bonds.[35] They were also nervous about piling up debt: a survey made in the early 1950s showed that only half of Americans believed that buying on time was a good thing.[36]

Thus, with the rising birth rate, and the move away from the central city to smaller suburban towns—and many of them at that time were small—there was the appearance, and probably to an extent the reality, of a movement toward a more family- and community-centered way of life. In 1960 70 percent of American households contained a breadwinner Dad, a homemaker Mom, and children.[37] (Today only 7 percent of households fit this pattern.)

Nonetheless, there is no evidence that the broad, long-term spread of the ethic of the self was reversed. Time spent with the entertainment media increased, especially after the advent of television in the early 1950s. Attitudes toward sexuality continued to loosen, in part due to the dislocations created by the war. Alcohol consumption increased: people were drinking about a third more liquor in the years immediately following the war than they had been immediately before.[38] Divorce rates were up, although once again a substantial proportion of these failed marriages were ones that had been hastily entered into during the war years.[39]

The general picture, then, for the immediate post-war years of the late 1940s and the 1950s was for the trend towards selfishness to be slowed, but not stopped by the great events of the years from 1929 to 1945. And as the Depression and war began to fade in the national memory, the self began once more to be a major preoccupation of Americans. The young people coming into the work force in the 1960s could not remember the Depression. They remembered instead the prosperity of the post-war period: the shady suburban streets, the fat hamburgers sizzling on backyard barbecues, the houses with recreation rooms, the new cars, the bureau drawers full of clothes. The prosperity their older brothers and sisters felt so grateful for seemed to them a norm. By 1962 some 62 percent of salaried or wage-earning heads of families had never been unemployed, and only 20 percent saw it as a threat.[40] By that year the number of families who owned stocks had doubled.[41] Two-thirds of American families were in debt, and half were carrying some sort of installment payments.[42] By this time, George Katona says, only the old, the disabled, non-whites, single mothers, and some farmers were poor. To be sure, as Michael Harrington and others began to point out at the time, these "invisible" poor constituted as much as a quarter of the population by some estimates.[43]

But by far the largest number of Americans were comfortably off. Prosperity was fifteen years old. A wartime hero, Dwight Eisenhower, had reigned as President for two terms, and almost eight years of peace. He was followed by John F. Kennedy, a bright young man with the future shining in his face. True, the shadow of the Cold War and nuclear weapons kept edging into the sunlight; but it really did seem to Americans that they could have whatever they wanted if they put their minds to it.

Unprecedented prosperity was the central fact about America in the roughly thirty years after the end of the war. It did not of itself produce the changes in attitude which were to come; but it made them possible. It was a necessary, if not sufficient, condition. For one thing, as should be obvious, the fact that there was plenty of money around allowed people to indulge themselves in ways that previous generations simply had not been able to afford. Perhaps more significantly, young people coming of age in the 1950s and 1960s faced fewer challenges, both personal and social, than other generations had. In the 1930s it had been difficult merely to find a job, much less build a career in business or a profession; in the post-war period there were jobs enough for nearly all, and careers to be had for the asking. Indeed, corporations which had in the past picked and chosen among an excess of applicants were now sending recruiters onto college campuses to persuade graduates to submit job applications.

But it was not just prosperity that made the United States in the 1950s and 1960s so comfortable a place to live in. Perhaps equally important was the fact that some of the social problems of an earlier day appeared to have been solved—or at least seemed to be on the way to solution.

In particular, the wave of immigration of the earlier years of the century had, for practical purposes, ended. Indeed, from 1931 to 1936 more aliens left the United States than came, for obvious reasons.[44] by the 1960s the children, grandchildren, and great-grandchildren of the immigrant wave had been, to a considerable extent, successfully integrated into American society. There still remained ethnic pockets where people spoke broken English and kept to the ways of the old country; but the ethnic slum, with its hollow-eyed children and despairing parents, was disappearing.

Taken as a whole, people with names like Weinberg and Dubrowski felt like Americans—felt they were no different from other Americans. They were beginning to make their way into the mainstream in business, professions, or the ordinary union jobs their families had done so much to create. Moreover, they were buying homes in the suburban communities where the Smiths, Browns, and Joneses had lived for two generations. Ethnic conflicts continued to exist, especially in the central cities, where kids from Catholic, Jewish, or other backgrounds taunted each other or fought in the streets. But to a considerable extent the descendants of the immigrants had stopped defining themselves primarily by their ethnicity. Yes, they might feel a certain tie to Italy or Poland, but they were beginning to feel kinships based more on neighborhood, community, trade, avocation, which crossed ethnic lines. It was significant that by the 1960s performers no longer felt it necessary to adopt Anglo-Saxon names, as in the cases of actors Ben Gazzara and Anthony Franciosa.

Even blacks, it appeared, were finally seeing some improvements. In the 1930s even well-known black entertainers like Louis Armstrong and Duke Ellington had difficulty getting rooms in downtown hotels; by the 1950s these performers were welcomed in restaurants and hotels their fathers could have entered only as porters and dishwashers.[45] In the 1930s there were no blacks in professional baseball outside the black leagues and few in other sports; by the 1950s blacks were coming to play a dominant role in sports. In the 1930s it was rare to see blacks in banks, department stores, groceries in white neighborhoods, even in the North; in the post-war era blacks were at times finding homes in white neighborhoods, although usually with some difficulty. Blacks were still, in the 1950s and 1960s, second-class citizens, in the main confined to the lower-paying jobs and living in run-down neighborhoods. But it was clear that things had gotten better for them, and it could be believed that finally the intractable black problem was on the way to being solved.

A kind of by-product of prosperity and the movement of ethnic groups into the mainstream was a general lessening of industrial turmoil. There were still strikes, still picket lines, still hard feelings; but the bitter warfare between management and labor that had existed in the 1930s and before had been tempered. Industrial conflict was more and more being fought out in conference rooms rather than on sidewalks.

I am by no means suggesting that America of the post-war years was a utopia: it was far from that, as unemployed blacks living in slum apartments, and elderly widows eating cat food to stretch out small pensions, knew. There was racism still, there was a quickening of the war in Viet Nam, and more. But *relatively* speaking—compared with what had gone before and what would come—post-war America was a prosperous and orderly place.

And it was certainly this order and prosperity which helped to create the phenomenon that characterizes post-war America more than any other—the sudden and astonishing arrival of television as a major force in American life.

The idea that pictures could be sent through the air dates back to the moment when "wireless telephony" proved to be feasible; much of the theory had been worked out in the early days of radio broadcasting. By the late 1920s television had proven practical, although there were still a good many problems to be solved, especially in regard to cost. NBC established an experimental station in 1928 and began transmitting from the Empire State Building in 1931. In that year, too, CBS began regular programming from its own experimental station. In 1939 NBC started a program of regular broadcasts from the World's Fair in New York, but the war held up further development of the medium.[46]

The first sponsored television broadcast was made in 1946 when Gillette, the razor company, sponsored the dramatic Joe Louis–Billy Conn boxing match.[47] For the moment expansion was slow: the American people were primarily concerned with buying the refrigerators and automobiles they had been unable to get during the war. In 1950 only 9 percent of American homes had television sets.[48]

The boom would certainly have come anyway, but it was triggered by the so-called Kefauver investigation of organized crime, which began in March 1950. (Estes Kefauver was a relatively obscure senator from Tennessee who suddenly found himself famous and a potential presidential candidate as a consequence of the televised hearings.) The hearings were held in the Foley Square Courthouse in New York, and the independent station WPIX decided to telecast them.[49] According to one man connected to the station, "We didn't expect much of a public response. Neither did the Committee."[50]

How wrong they were. Very quickly the word spread. Here was a chance to see in action what were supposed to be big-time criminals. People hurried out to buy TV sets expressly to watch the hearings. The audience grew as other stations began picking up the WPIX broadcasts, until it reached 30 million people. Movie houses were empty, and one Chicago department store advertised a 10 percent discount during the hours the hearings were on.[51] Suddenly, all in a moment, TV was no longer a luxury; it was now an essential item in every American home. Four years after the Kefauver hearings there were television sets in over half of American homes; by 1960 it was 88 percent.[52] TV ownership increased every year thereafter until 1974, when it leveled off at 97 percent for a year, due mainly to the slowing of postwar prosperity. Even as late as 1972 and 1973, long after the medium had saturated the country, the industry was selling 17 million sets a year, either to replace outmoded ones or to add one to a bedroom or den.[53] By 1970 almost a third of American homes had more than one television set; and 11 percent had three or more.[54]

As the number of television homes increased, so did the hours the sets were on. In 1950 people who owned television sets had them on for just over four and a half hours a day; by 1970 it was almost six hours a day.[55] These figures, it should be noted, are averages for the year; winter viewing in many families was over nine hours a day. (As of 1987 that breaks down to 4:19 hours for adults, 3:02 for teenagers, and 3:01 for children.) Today 98 percent of homes have television, 96 percent of them color television, 60 percent two or more sets, and the majority either cable, a VCR, or both.[56]

Few changes to American society have come with such astonishing swiftness. An activity that barely existed in 1950, within ten years became a major preoccupation of the vast majority of Americans. Forty-year-olds who in 1950 had never seen a television show in their lives found themselves, within a year or two, devoting the bulk of their leisure time to it. Erik Barnouw reported that, in the early 1950s, "Juke-box receipts were down. Public libraries, including the New York Public Library, reported a drop in book circulation, and many stores reported sales down. Radio listening was off in television cities: the Bob Hope rating dropped from 23.8 in 1949 to 12.7 in 1951 and continued downward."[57] By 1951 movie attendance in cities with television was off some 20 to 40 percent, while it remained high, or even increased, in cities without television.[58] It was a phenomenon—a major shift in behavior patterns.

Attention to the set was not invariably rapt. Studies showed that for perhaps 20 percent of the time the TV set was playing to an empty room.[59] Furthermore, people frequently did other things while watching—ironed clothes, ate, conversed, drank, cooked, made love. But unlike today's radio, television was never merely a backdrop to other activities: it was the essential thing, with the other activities secondary. And, because it drew people deeper into the media, it contributed substantially to the fragmentation of American life which the focus on the self had engendered. Although people did, of course, watch television in groups, it was essentially a private occupation, a way of disengaging the self from others—a wholly in-taking occupation.

It is therefore not surprising that it slotted easily into the neat, prosperous, self-involved lives that most Americans were living. And as part of that easy, smooth life, it contributed to what happened next. For those who had seen the Depression and the war, neatness and smoothness were fine; for those who had seen nothing but clean suburban houses, close-clipped suburban lawns, Sunday afternoon barbecues, Monday night football, it seemed bland and sterile.

How much a wish for a more vigorous, textured life produced new strains in the society is difficult to know; but the effect, certainly, was there. Right from the earliest of the post-war years, two movements that would have profound effects on the culture began to appear out of the soil. One was a gradual loosening of the strictures against pornography, the major residue of the Victorian age, which had demonstrated considerable staying power. The second was the intellectual and literary movement known as "the Beats."

CHAPTER 15

The Triumph of Selfishness

As we have seen, one of the key aspects of Victorianism was the unremitting drive mounted against the depiction of sexuality in almost all forms. As Victorians saw it, sex was a dangerous appetite that could be easily aroused. As a consequence erotic materials should be kept from the most easily arousable, who were believed to be children and most especially working-class males, which meant in effect that the materials had to be generally proscribed.

There was, however, one exception to the obscenity rules: works of "art," however that term might be defined, were exempt. This, too, followed logically from Victorian belief: it was understood that middle class men and women had sufficient self-control not to be thrown all of a heap by the sight of a female breast, as for example those of the Botticelli Venus. But this kind of self-control could not be expected of the half-formed young, or of working people who were seen as easily swayed by emotion to riot and rape. In the United States this feeling was exacerbated by the presence in the midst of the society of a huge immigrant mass who were not Victorians and did indeed seem to be overly sexual creatures. Art, however, was something, it was assumed, of not much interest to the young or working people. They were unlikely to haunt museums or purchase volumes of reproductions of Rubens nudes.

This was not unwritten law, but was spelled out, as we have seen, by Sir Alexander Cockburn. His 1868 opinion—which became the basic rule—said, "The test of obscenity is this, whether the tendency of the matter charged as obscene is to deprave and corrupt those whose minds are open to such immoral influences, and into whose hands a publication of this sort may fall."[1] It was not the nature of the obscene matter, but the audience that mattered: a folio edition of Rubens was not likely to fall into the hands of those it might corrupt; a set of cheap sketches of nude women turned out by a penny press was.

So long as jurists were concerned with the audience and not the work, the Cockburn rule would hold up. Through the second half of the Victorian Age the attack on obscenity was unrelenting and almost entirely successful. Sex simply was not mentioned anywhere in anything that was allowed to float down the cultural mainstream of America. But the realist writers like Dreiser, the painters of the Ashcan School, the photographers of the mop and pail brigade, determined as they were to depict the "more real" underside of life, were bound to insist that they should be able to deal with human sexuality. In the 1920s particularly, the war against censorship became a major issue for intellectuals and bohemians. And as the battle mounted, without anyone's being quite aware of it, attention began to switch from the potential audience for obscenity to the object itself. And now the exception made for art opened a loophole in the Cockburn rule big enough to drive a truck through.

A halfway stage in the change of interpretation came in 1933 with Judge John Woolsey's ruling that James Joyce's *Ulysses* was not obscene. Woolsey believed that, although the book contained sexual and scatological references, and many four-letter words, he did not "detect the leer of the sensualist" in it. The book would not rouse lustful feelings, despite "its unusual frankness."[2] Woolsey was still concerned about the possibility of heightened passions, but he no longer was worried about whether the book might fall into the hands of children and working people. And from this point on the censors were fighting a losing battle. "The general rule of thumb became that if a book contained some social merit, and was not clearly directed toward arousing sexual passions, it could not be held obscene."[3] The Supreme Court was now saying, "the portrayal of sex, for example, in art and literature, and scientific works, is not in itself sufficient reason to deny material the constitutional protection of freedom of speech and press."[4]

This idea that social merit would excuse obscenity, however defined, was the formula under which *Esquire* magazine was run from 1933. Looked back on from a day when magazines routinely show vaginal labia, the furor over *Esquire* seems ludicrous. In the 1930s the magazine occasionally carried rather vague sketches that included bare breasts, but many issues would appear with not even this amount of nudity; photographs at most showed women in one-piece bathing suits, a sight anybody could see on any beach at the time. Nonetheless, the magazine, which did run a good many suggestive cartoons, was celebrated for its boldness, and was not generally found lying on coffee tables in middle-class homes. But *Esquire*'s notable editor Arnold Gingrich knew what he was doing, for he surrounded the hints of sex with fiction and poetry by Ezra Pound, F. Scott Fitzgerald, Ernest Hemingway, and other literary lights, packaged the whole thing expensively, and generally gave the magazine a certain intellectual tone.

In 1943 the Post Office Department decided that an intellectual tone was not guise enough to send bare breasts through the mail. The Post Office Department had long been authorized to withhold obscene materials from the mails. This mattered to magazine publishers more than it would appear; it was—and is—not merely subscriptions which are mailed. So cheap is the postal rate under the so-called Second Class Entry that entire issues of magazines are "mailed" to distribution warehouses around the country. It is in fact a subsidy originally intended by the United States government to promote the circulation of newspapers and other periodicals in the interest of creating an informed public. It is today available to any periodical that comes out regularly.[5]

Losing the Second Class Entry can be expensive, even fatal to a magazine. The Postmaster-General's claim was not that *Esquire* was obscene but that the intention of Congress when it defined the Second Class privilege was to give the subsidy only to magazines that "contribute to the public good and the public welfare." The Supreme Court's response was that, "To permit a postmaster general to decide what contributes to the public good . . . would be to grant the postmaster general a power of censorship."[6]

But the Post Office Department was hardly the only censoring body in the United States. All across the country in thousands of villages, towns, counties, and cities there existed a huge jumble of obscenity statutes and pornography ordinances which had been pushed through by aldermen, town selectmen, and other authorities. These rules were being enforced by thousands of county sheriffs, police chiefs, and other officers of the law, often goaded into action by ministers, local purity groups, and the Catholic Legion of Decency. These lawmen, guided either by what lay in their own hearts, or the admonitions of religious and other groups, frequently pulled books and magazines they deemed offensive from the racks in bookstores and newsstands. Publishers had little recourse. A publisher could of course sue, and usually win; but this was of little use to a magazine publisher, for by the time the case was called, the issues which had been banned would be long dead. A book publisher had a longer sales period to work with, but in most instances it was simply not worth fighting sporadic cases of censorship in small towns here and there.

The net result was that publishers of the period were very careful to observe unwritten, but well-understood limits assumed to represent the general view. A magazine like the present *Playboy*, much less *Hustler*, simply could not have been published in the 1950s, or even the 1960s. The lines were quite fine-drawn: you could, for example, show any part of the naked female breast but the nipple. This rule was waived for art magazines, however, but even these magazines could not show pubic hair, and art directors on the staffs of such magazines grew quite adept at airbrushing, to leave a strange patch of white in the appropriate area.

Feelings about this whole question of censorship in the post-war period were quite intense. On the one hand, churches and various guardians of the public morality were determined to keep the old barriers up; on the other hand, the publishers, writers, and artists were determined to tear them down. The critical moment came in 1957, with the landmark Roth case, in which the Supreme Court came down in favor of the publisher. By this time the concern for "those whose minds are open to such immoral influences" was gone. The judges were interested solely in the nature of the material deemed obscene, and this confronted them with a problem: if breasts on, say, Ingres's Vénus Anadyoméne were not obscene, how could a picture of the same organs on a nightclub stripper be held illegal? The Supreme Court waffled. The key provision in its decision was "whether to the average person applying contemporary community standards the dominant theme of the material taken as a whole appeals to the prurient interest."[7] Two years later, working from these dicta, the Court found that D. H. Lawrence's *Lady Chatterley's Lover* was not obscene. The ball game, essentially, was over.

Whatever the Supreme Court thought it was doing in the Roth ruling, it did in fact open the gates. For one thing, "community standards" was an extremely slippery rule to catch by the tail, as standards vary considerably not only between communities but within communities. On the other hand, "the dominant theme" and "taken as a whole" are relatively easy to measure. From this point on it became difficult to prosecute obscenity cases unless the material made no pretense of being anything else. Publishers were not that stupid. It was perfectly obvious that a book with a few red-hot seduction scenes which was otherwise about life in a nunnery could not be banned, even though the seductive scenes were the selling point. At this point the whole thing became an exercise in seeing how far the limits could be pressed.

One important factor in helping to liberalize American attitudes toward the display of sex was the publication of the so-called Kinsey Reports—the volume on the male in 1948, the one on the female in 1953. They were issued by a Philadelphia medical publishing house, W. B. Saunders, which found itself, vastly to its astonishment, with an enormous seller on its hands when the first volume appeared. It was reviewed everywhere, praised, attacked, mulled over by pundits. It made Kinsey a famous name, and, inevitably, the subject of a flock of jokes. The volume on women caused an even greater furor, with thousands of journalists attempting to get advance copies, and stories appearing everywhere for months. Both books had huge sales for what were basically medical texts, although the numbers of people who actually waded through the hundreds of pages of charts and graphs that were the essence of them were undoubtedly small. Nonetheless, as a consequence of the Kinsey Reports, readers of both highbrow and popular publications found themselves immersed in long discussions of subjects like oral sex and juvenile homosexuality, which many barely knew the existence of, much less given any serious thought to.

The effect of the Kinsey Reports, then, was suddenly to make permissible the public discussion of sex in all its variety. In particular, the highly competitive commercial magazine industry, always looking for enticing new subject matter, found that it could now cover forbidden topics, like sex in the colleges, adultery, abortion, if they were presented as serious reports, replete with statistics drawn from scholarly studies, and quotes from university professors. They were aided in so doing by an increasing number of studies of sex which the Kinsey work triggered. By the late 1950s magazines were routinely doing stories with titles like "A Doctor Talks About Frigidity," "Memories That Inhibit Married Love," and "What Keeps a Husband Faithful,"[8] subjects that would not have been covered ten years earlier.

The effect of the *Esquire*, Roth, and *Lady Chatterley's Lover* cases and the publication of the Kinsey Reports had made it abundantly evident that a lot of previously censorable material could now be published, provided it was at least wrapped in some sort of pseudo-scientific or arty mantle. It was certainly with this in mind that a young man named Hugh Hefner decided to start a magazine that would have a substantial impact on American morality. Hefner had grown up a fan of *Esquire*, attracted, as were many young men of his generation, by the combination of sex and presumed sophistication the magazine offered. He landed a job in the promotion department of the magazine, and when he was refused a five-dollar raise he quit[9] and went back to Chicago to start his own magazine, using the *Esquire* formula of sex wrapped up in an intellectual package. Scraping together what today seems like a ludicrously small amount of money, in 1953 he managed to put together a fairly crude first issue of *Playboy*, which he sent onto the newsstands undated, in case he had to leave it there for more than a month. The magazine featured Marilyn Monroe on the cover, then only beginning to be widely celebrated, and inside a nude photograph of her, which showed her breasts and a considerable expanse of thigh, but little else of moment. With the Monroe picture as the primary lure the issue sold out a printing of 70,000 copies and Hefner was on his way.[10]

By comparison with what is routinely shown in magazines today, the early issues of *Playboy* are astonishingly tame. The main advance over what competing magazines were doing was to show the full female breast. There were also some fairly bawdy but hardly pornographic stories, off-color jokes, and the like, but nothing that could not have been found in many hardcover books of the day. Nonetheless, *Playboy* was considered an exceedingly steamy magazine; as had been the case with *Esquire,* middle-class families would not have displayed a copy on the coffee table, and it would be almost a decade before the New York Public Library's research division considered it worthy of inclusion in its stacks.

The great success of *Playboy* inevitably produced a flock of imitators through the mid-to-late 1950s, among the earliest of them being *Escapade, Nugget, Dude, Gent,* and eventually *Playboy's* major challenger, *Penthouse.* Some of these magazines were more daring than *Playboy,* some of them less, but Hefner was the first to persuade major advertisers to come into the magazine in large numbers. Ads for liquor, automobiles, and name-brand clothes conferred a legitimacy on the magazine that the others could not acquire, and *Playboy* went on to dominate the field for some time. Paul Gebhard of the Kinsey Institute said, "Hefner's genius was to associate sex with upward mobility." [11]

The competition, however, forced Hefner to test the limits constantly. In June 1957 *Playboy* published a picture showing a few inches of the buttock cleavage of a young woman. It occasioned no particular uproar, and by the next year the magazine was routinely showing the bare rumps of its subjects. What is significant, however, is not what *Playboy* did show, but what it did not. The frontier ahead was what came to be called full frontal nudity, that is to say, the showing of female pubic hair. This had always been impermissible: even Botticelli and Michelangelo understood that they could not cross this line. Despite the liberality growing through the society in the 1950s and the 1960s, despite the Roth and *Lady Chatterley's Lover* cases, Hefner sensed that his middle-class readers and the advertisers courting them were not quite ready. He may have been unduly cautious, but, as we shall see, probably not. Not until 1969, when promiscuity as an ideal had already been proclaimed by the flower children, and the movement was peaking in the Woodstock celebration, would *Playboy* show pubic hair, and its debut in this field was a very ambiguous strobographic shot of a woman in motion. The real breakthrough came in the 1970s, when the showing of pubic hair became routine. The April 1974 issue carried a picture of Marilyn Chambers in a tub of water with her hands placed near her vagina in such a way as to suggest that she might be doing something down there, and as the decade wore on *Playboy* and its competitors were beginning to show pictures of female sex organs at close range, and the sex act itself. The really sweeping advances, thus, came only after 1970.

One more element in the mix that helped to liberalize American attitudes toward sex was the sex education movement of the 1960s. The idea that children should be taught about sex in the schools dates back to the vice reports which appeared in such abundance in the first decade of the twentieth century. Many of them called for sex education.

Furthermore, the concept of Dewey and his followers that schools should teach life-enhancement certainly encouraged schools to add sex courses to their curricula. In fact, very little that could be defined as sex education was instituted in most schools.[12] A few progressive private schools tried it; and in some others the school nurse might run a film on menstruation for the girls, usually well after most of them had begun menstruating, and a coach might give the boys a talk on the evils of playing with yourself, again many years after most of the boys had begun masturbating. Right through the century, into the 1960s, sex education, practically speaking, did not exist for most students.

But by the 1960s, with what appeared to be increasing amounts of sexual activity among the young, along with rising rates of premarital pregnancy and venereal disease, pressure for sex education increased. It was argued, among other things, that students were already getting a considerable dose of sex education from *Playboy* and its imitators, and that there was little point in trying to keep from them the news that babies were not found in cabbage patches.

One educator who was particularly concerned was Paul Cook, school superintendent in Anaheim, in Orange County, California.[13] The town had become something of a way-station for people coming in from elsewhere in the United States, especially the South and the Midwest, who then moved on to other places in California. There was a good deal of dislocation among the newcomers, with a lot of divorce, and a high rate of teenage pregnancy among the girls, which led to an increasing drop-out rate. Cook reasoned that, if premarital pregnancy could be reduced, the drop-out rate would fall. In 1963 he surveyed the local parents, and discovered that an astonishing 90 percent favored instituting a sex education program in the schools. With this as ammunition, he consulted Lester A. Kirkendall, one of the most respected figures in sex education theory of the time, as well as others, and put in a comprehensive sex education program. The Anaheim program was carefully studied by interested people and became a model for sex education programs elsewhere.

Anaheim, however, was a rather odd place for this bold experiment in sex education to take place. Orange County is a bastion of conservatism—the John Birch Society was headquartered there—and Anaheim possessed a very vocal conservative minority which was supported by a powerful local newspaper. Within a few years groups opposed to sex education began an attack on the program, with the newspaper leading the fight. "They printed all kinds of things that weren't true," said Sally Williams, a school nurse who was coordinator of the program, "like we were demonstrating sex in the class, or ran movies with rape scenes in them. It was very hard to fight that, because the paper wouldn't run our side of it, only the anti side."[14] In the end the antis won, primarily because parents who supported the program—far and away the majority—were intimidated into not speaking out. Williams was taken off the program, Paul Cook went into retirement, and the program was wound down.

What happened in Anaheim happened in many places across the United States in a lesser way. Through the mid-1960s scores of sex education programs were conceived, and many started. A number of publishers, smelling a boom in sex education on the way, started to put together packages of materials. But by the end of the decade the anti-sex education forces had rallied. Community school boards backed off, and the sex education movement was effectively killed before it had really gotten started. By 1980, after fifteen years of effort, only about 10 percent of school students were getting "a comprehensive sex education program," according to Peter Scales, then a research analyst studying sex education programs for the federal government.[15]

But despite the fact that sex education programs never got off the ground in most school systems and crashed, where they actually flew in other places, the subject got a lot of attention and was discussed both in the press and by thousands of American school boards, P.T.A.s, and other related organizations. Ordinary parents, teachers, school administrators, and school board members found themselves talking in public about masturbation, condoms, and similar subjects, which earlier on they might not even have mentioned to their spouses. The people active in sex education issues were a minority, and probably a small one at that; but, according to a Gallup poll, some two-thirds of Americans had always approved of the idea of sex education, going back into the 1940s.[16] Their reasons were, as always, multifarious: some wanted the schools to take off their backs a job they knew they should be doing, but were not; others were hoping that sex education would help control the sexual activity of the young; still others saw it as curing social problems, like venereal disease and teenage pregnancy. But many of these parents had come genuinely to believe that sex was life-enhancing, and that people ought to be showed how to make it work for them instead of against them. In sum, the sex education movement, especially in the 1960s, helped to promote the idea that sex ought to be a joy for all.

When we stand back and look at what I have chosen to call the post-war period—from the late 1940s to the early 1970s—we can see running through it a new openness about sex that had been not merely unpalatable but actually illegal even ten years before. Much of what was said and shown in the few sex education texts that did appear in the 1960s would have been banned from the mails by the postal authorities in the 1930s. But now, in the years after the war, there were the Kinsey Reports saying that the things people thought they were alone in doing were in fact common practice; *Playboy* and its imitators insisting that sex ought to be fun for all; distinguished college professors telling

P.T.A.s that it was all right to talk to kids in school rooms about oral sex

The consequence was that attitudes towards sex held by millions of ordinary people were rapidly changing. They were coming to believe that sex had an important, perhaps central, role in human life; coming to believe that a lot of what they had been raised to think were perversions was commonly practiced in the marriage bed; and, perhaps most important, coming to see that sex was a permissible subject for art, for the popular press, for conversation even in schools and church meeting rooms. (Many Protestant ministers in particular were in the forefront of the sex education fight.) Intellectuals, artists, and social scientists had espoused this view for decades; but now ideas that once had been discussed only in bohemian coffeehouses and social sciences labs were common property. As we shall eventually see, sex *behavior* was changing only slowly through this period; but a remarkable shift in *attitude* was taking place at an increasing pace, through the 1950s and 1960s.

It is against this backdrop that we must watch rise into sight what appeared to be a literary movement but was in fact most significant as a way of life. How important the literature created by the Beats, or beatniks, as they came to be called, will prove in the long run is an open question; but there can be no doubt that their influence on American behavior was immense, particularly in view of how tiny a group they were at the beginning.

The thinking of the Beats was not radically different from that of the bohemians of earlier times: it was mainly that they carried it all a step—or several steps—farther. The original group coalesced at Columbia University in the winter of 1943–44 and the next year or so. According to Tom Clark in his biography of Jack Kerouac, they included at first Allen Ginsberg, William Burroughs, Kerouac, and eventually Neal Cassady, who would be seen by the others as the archetypical Beat.[17] Around this core there grew a larger group with a shifting personnel that came and went as they traveled nervously around the country to visit little pockets of like-minded people living in shacks, seedy apartments, and flophouses in San Francisco, Mexico, Texas, New Orleans, Florida, and New York.[18] Many, if not most, of them had literary ambitions. They knew the legends of the literary group which had arisen in the United States after the previous war, which had thrown into fame Ernest Hemingway, Edna St. Vincent Millay, Scott Fitzgerald, William Faulkner, Malcolm Cowley, Edmund Wilson, and more. They fed on the bohemianism of the 1920s, which saw the artist as an outsider whose duty to his art required him to strip away the smug pretensions of the bourgeosie in order to expose the raw beauty of real life. They simply took it for granted that they stood in opposition to society, prepared to fight it out for the rights of individuals. They *began* by believing that the self came first.

Kerouac is generally credited with entitling them a "Beat generation," and in 1952 the term was given some currency by Clellon Holmes,
who was in the group but not quite of it, in an article for the *New York
Sunday Times Magazine*[19] which discussed the Beats as a phenomenon
of the post-war period. Holmes said, "More than mere weariness, it
implies the feeling of having been used, of being raw. It involved a sort
of nakedness of the mind, and, ultimately, of soul, a feeling of being
reduced to the bedrock of consciousness. In short, it means being un-
dramatically pushed up against the wall of oneself. A man is best whenever he goes for broke and wagers the sum of his resources on a single
number, and the young generation has done that continually from
youth."[20]

Kerouac offered a similar thought: The Beats "were like the man
with the dungeon stone and the gloom, rising from the underground,
the sordid hipsters of America, a new beat generation."[21] Stripped of
romanticism, Holmes said, "the philosophy is that of young hedonists
who don't really care whether something is good or evil, as long as it is
enjoyable."[22]

The original Beats were a strange mix: Lucien Carr and William
Burroughs—who was older than the others—were from comfortably-
off St. Louis families able to keep them afloat with small incomes for
periods. Jack Kerouac was from a working-class family and went to
Columbia thanks to a football scholarship. Allen Ginsberg was the son
of Jewish immigrants with strong left-wing and literary inclinations—
his father, Louis Ginsberg, was a respected poet. Neal Cassady had been
abandoned by his mother, to be raised by an alcoholic father, or really,
on the streets, where he became an accomplished automobile thief at
an early age.[23]

What they had in common was a clear sense that they were outsiders;
indeed, they seemed to pursue deviance as a goal. A high percentage
of them were bisexual; all of them routinely drank heavily and were
using vast quantities of drugs, even as students, at a time when this was
not done. Few of them ever established themselves in regular, ongoing
jobs, much less professions; most of them would at one time or another
be jailed for criminal offenses—Lucien Carr killed a man who had been
making homosexual demands on him,[24] and Burroughs accidentally
killed his wife during a game with guns;[25] and many of them, including Ginsberg and Kerouac, spent time in mental institutions.[26] It will
be argued that these people were committed not because they were
insane but because they were deviant, and there is no doubt a degree
of truth in this; nonetheless, judged on any ordinary basis, the behavior
of the core group of Beats, the ones who established the model, exhibited a great deal of psychopathology.

It is important to what happened later to see that these early Beats
believed deeply in sexual freedom. Allen Ginsberg, if his sympathetic

biographer Barry Miles is to be believed, was obsessed with sex. More-over, they were committed to the idea that through drugs they could find the transcendent insight they were looking for. Ginsberg, again, made his search for the drug that would open visionary doors—for him a pilgrimage. For fifteen years, until he changed his views in 1963, "he took every powerful hallucinogen he could find, from laughing gas to mescaline; he sniffed ether and shot heroin . . . He virtually ignored everyday life in his concentrated effort to widen his consciousness."[27]

Although Allen Ginsberg did eventually conclude that he had a mes-sage for the world, which he tried to impart to listeners especially after he became a folk hero in the 1960s, at bottom the Beat movement was not political or activist. Barry Miles says, "They rejected the conformist, consumer society of America in the late forties, but were not thinking in terms of rebelling against it. The Beat Generation began with per-sonal exploration." And he quotes Ginsberg as saying of it, "It wasn't a political or social rebellion. Everybody had some form of break in their consciousness or an experience or a taste of a larger consciousness or satori."[28]

Basic to their thinking was the idea that through drugs or other means they ought to be able to enlarge their consciousness. It has never been entirely clear what they expected this experience to be, except that it would lead to peering deeper into things, or seeing connections that most people miss.

They also tended to identify with various types of outsiders, or de-viants, such as drug addicts, petty criminals, beboppers and their hip-ster hangers-on. Some of them spent a lot of time in the Times Square area cultivating such people, and William Burroughs became involved to the extent that he was acting as a receiver of stolen goods.[29] They believed that "holy madmen," and "angel-headed hipsters" had a way of managing life that might teach them something.

This interest in people outside of society, coupled with the idea of the transcendent experience as the door not merely to knowledge, but to a new way of life, inevitably brought the Beats to focus on the self. The social system was almost irrelevant to them; it was the individual, not the group, experience that was of paramount importance. The at-titude was expressed in Ginsberg's *Howl*, where "angelheaded hipsters" burned for "the ancient heavenly connection."[30]

Another key idea was a belief in spontaneity in art and life. Kerouac and some of the others at times tried to write spontaneously, putting down on paper whatever came into their heads. In life they demon-strated spontaneity by partying for days at a stretch or jumping into cars on a whim and driving across the country to see a friend. Again, spontaneity is by definition an expression of self, not the group.

Lucien Carr said, "In those years at Columbia, we really did have something going. It was a rebellious group, I suppose, of which there are many on campuses, but it was one that really was dedicated to a 'New Vision'. . . it was trying to look at the world in a new light, trying to look at the world in a way that gave it some meaning. . . And it was through Jack and Allen, principally, that it was going to be done."[31]

The Beats were not professional philosophers, or even rigorous thinkers attempting to put together systematic and consistent sets of ideas. They would, in fact, have seen such an attempt as constricting, contrary to the spontaneity they thought essential to life. It is therefore difficult to extract from their work a basic philosophy.

But that same writing—especially the semi-autobiographic books of Kerouac—makes it quite clear that they saw their own lives as prescriptive. Running through this body of work was the belief that we, the Beats, are the angelheaded hipsters who are living hard and real.

And in their behavior they were exhibiting a dedication to the self that was all-consuming. Lovers were abandoned when there was anything better to do; children were ignored, even disowned, for years at a stretch; friendships rapidly wore thin as the needs of opposite selves rubbed together; groups merged and splintered.

There were exceptions: Allen Ginsberg, in particular, appears to have been capable of sustaining longer-term friendships and pair relationships. But that was not typical of the Beats.

The Beats would say in their defense that they were not interested in stable relationships, an orderly world, or even a sane one; they were at the edge of something, pushing the horizons of life further, testing the limits of human possibility, looking deeper into things than did the people of the square world of the suburbs with their automatic jobs and their automated pleasures. There is a certain validity to this defense. Norman Mailer, in his famous piece on the white Negro, says that the task is to "divorce oneself from society, to exist without roots, to set out on that uncharted journey into the rebellious imperatives of self. In short, whether the life is criminal or not, the decision is to encourage the psychopath in oneself."[32] Richard Hofstadter, who was not a supporter of the movement, says, "The type of alienation represented by the beatniks is, in their own terms, disaffiliated. They have walked out on the world of the squares . . . have repudiated the path of intellectualism and have committed themselves to a life of sensation."[33]

What the Beats had done, then, was to take the ethic of the self which had grown up in the early decades of the century, and driven it to its furthest extremes. There were no limits: anything that suited the self was permitted, no matter how it might impinge upon others. For the Beats, self-denial was the primary sin.

Looked at from a distance, however, the Beats seem more like lost children than geniuses burning with a passion for life. The word "sad" echoes through the key book of the movement, *On the Road*, like a distant bell constantly tolling. They floundered and wandered and eventually began to flicker and go out. Neal Cassady died of congestive heart failure in his forties from a combination of drugs and liquor;[34] and Kerouac, the spokesman for the group, also died in his forties when his liver gave out from alcoholism.[35] He had been told that he must stop drinking or die, and he chose to die.

The Beats, then, were deeply committed to both a philosophy and a lifestyle of extreme deviance. They were not so much antisocial but asocial—without any large concern for any but their own individual selves. In the 1952 *New York Times* piece Clellon Holmes speaks specifically of the "absence of personal and social values"[36] in their outlooks. It was a philosophy of ingestion, of personal sensation, of the immediate moment. And yet on this deviant philosophy, created by a group haunted by serious psychopathology, was built a creed which swept through a younger generation in the 1960s and left marks on the society that are still evident today.

By the time of the Clellon Holmes article in 1952 the basic group had been together in one form or another for some eight years. They had been for some time passing their novels and poems around among themselves, and in the bars and cold-water flats of Greenwich Village and similar places, and some of them had become known as undiscovered geniuses whose time was coming. By the late 1940s Kerouac had come to the attention of Malcolm Cowley.[37] who had discovered John Cheever, resurrected the reputation of William Faulkner, and helped to build the reputation of friends like Hart Crane, John Dos Passos, e. e. cummings, and others. Cowley was a powerful force in the literary world, and set about trying to get Kerouac published.

By the mid-1950s some of the group had come together in San Francisco. A series of poetry readings there, with Ginsberg usually offering an impassioned reading of *Howl*, made them locally famous.[38] In 1957 Kerouac's *On the Road* was published, and almost at the same moment *Howl* was subject to a highly publicized banning. The Beats were exceedingly good copy, and Ginsberg, who had worked in advertising for short periods, proved a good publicist. Over the next few months there were articles in a dozen different magazines, including *Commentary, The Library Journal, Harper's, Newsweek, Vogue,*[39] and *Time,* which said of Kerouac, "with his barbaric yawp of a book, Kerouac commands attention as a kind of literary James Dean."[40] There were appearances on talk shows, newspaper interviews, press parties and the whole uproar which follows the arrival of what appears to be a new comet. Tom Clark said, "Without asking [Kerouac] the national media had appointed him the spokesman for an entire generation, and his slightest thoughts made the wires hum coast to coast."[41] Kerouac, as much as he had felt bitterly neglected previously, could not handle the sudden celebrity. He began drinking more than ever, became reclusive, and finally settled in to live with his mother, who tried to keep his old friends from seeing him. In the end Allen Ginsberg became the spokesman for the movement.

Meanwhile, all over America aspiring young writers, aspiring bohemians, and just plain aspirants were seeing that a new genius, or group of geniuses, with new ideas and an exciting lifestyle, had popped into view. *On the Road* sold 500,000 copies in paperback in the first two years after its publication.[42] It was followed by a spate of books by Kerouac and others of the group, rushed out by publishers who were afraid that the Beat movement would prove to be temporary. In 1960 *The Subterraneans* was filmed with Leslie Caron and George Peppard in the lead and some important jazzmen, including Gerry Mulligan, in the film. Ginsberg's *Howl* became the best-known poem of the day. William Burroughs was finally able to get his work published, and some of the others, such as Gregory Corso, broke into print, too. From this work, and especially from the press accounts, millions of young Americans learned about holy madness, spontaneity in life and art, the anti-bourgeois virtues of getting high and staying high.

The connection between the Beats of the 1950s and the hippies of the 1960s was direct. In fact, these were not two separate movements: the second was merely an extension of the first, taking over intact much of the philosophy of spontaneity, sensation, the restless hoboing around the country and much else. As Morris Dickstein said in his study of the counter-culture, "The immediate underpinnings of the sixties appear in the dissident works of the late fifties."[43] A Kerouac biographer, Gerald Nicosia says that the author's *The Dharma Bums* "served many [hippies] as a survival manual."[44] Virtually every important aspect of the hippie movement had been tried out by the Beats first. Allen Ginsberg was deeply involved with Timothy Leary's experiments with mind-altering drugs as early as 1960, long before Leary was known outside of a small circle, and he brought Kerouac and some of the others in.[45] Ginsberg had made long visits to South America, where he had tried a number of obscure Indian drugs, and to India, where he learned a good deal about the religions of the country; he was responsible for popularizing in the United States the famous "Hare Krishna," which later became so familiar a sound on western city streets.[46] Ginsberg, too, was proclaiming the philosophy of love which so pervaded the later groups.[47] Neal Cassady, the primal beatnik, was taken up by Ken Kesey and for a period lived in Kesey's psychedelic commune.[48] Ginsberg and many of the others were involved in the anti-Vietnam movement in the early days. And the Beats, especially, were advocating the holy joys of indiscriminate sexuality almost from the beginning. The slogan "Turn on, tune in, drop out" was an exact statement of the beatnik philosophy as it existed in the 1950s.

In 1963 Ginsberg, as a consequence of an insight he had after his long sojourn to India, Japan, and other eastern countries, changed his opinion of the need for the expanded consciousness. "I was suddenly free to love myself again, and therefore love the people around me, in the form that they already were."[49] He began bringing the message of love to his audiences in his readings and other public appearances. In so doing he was breaking with the position of the original Beats. None of the others had seen themselves as spokesmen for the Beat philosophy; indeed, Kerouac and Burroughs had become virtually recluses. By default Ginsberg, the sexual obsessive who had immersed himself in drugs for fifteen years, was left to carry the message on to the next generation. He became, as Dickstein has said, "the guru to a new generation."[50] I mean this as no denigration of Allen Ginsberg; he was intelligent, serious, and sincere, and he had struggled hard, often at considerable personal sacrifice, to find a new way for himself, and perhaps humanity. But there is no gainsaying that he felt that total sexual freedom and a proper use of drugs could be useful tools in developing that new awareness he hoped to bring to people.

Ginsberg's new approach was interesting. In the left-wing politics of his father, which had dominated intellectual political thought from the early days of the century, the basic idea had been to improve things for the individual by changing society. The new idea being forged by Ginsberg and others was the reverse: society would be improved by changing the individuals who made it up. It seemed obvious that if everybody learned to love everybody else, learned not to "lay their trips" on other people, learned not to "hassle" each other the social system would improve all in a stroke. Social progress, as well as art and individual salvation, would begin with the *self*. As the message came across to young people, it was turned into an almost irresistible fantasy: youth could transform the world by getting high and having sex. It was an offer that a lot of young people could not refuse.

This line of descent, from the Beats to the hippies, was recognized quite early. In the 1960s the *Chicago Tribune* accused Kerouac of having spawned "a deluge of alienated radicals, war protesters, dropouts, hippies, and even beats."[51] Ironically, Kerouac, who had a strong streak of blue-collar conservatism in his make-up, disliked the anti-war demonstrations and other aspects of the 1960s and was annoyed when his daughter, whom he had refused to acknowledge for years, turned up with "a hippie boyfriend."[52]

The rapid evolution of the Beat movement into the Woodstock nation which took place in the mid-1960s could not have occurred had not the soil been prepared by the new openness about sex brought about by the Kinsey Reports, *Playboy* and its followers, the sex education movement, the Roth and other Supreme Court decisions, and the increasing willingness of the mainstream media to discuss sex and related subjects. Had the morality of the immediate post-war years still obtained, it is unlikely that either *On the Road* or *Howl* could have been openly published, and it is certain that experimenters with drugs would have faced much stiffer punishments than they usually received. Perhaps more important, the media would have been far less likely to give the Beats as much attention as they did, and this of course was crucial for capturing the attention of the young who went on to form the Woodstock nation.

It is important at this point for us to make clear the distinction between the hippies and other strains in the counterculture of the 1960s, in particular the civil rights and anti-war movements, as well as environmentalism, vegetarianism, and other reforms and causes. While the populations concerned with this shopping basket of issues to some extent overlapped, and while people who demonstrated against the Vietnam war were also likely to be supporters of black rights, they were not the same populations. The anti-war movement included a lot of people from the suburban middle class, among them housewives, lawyers, executives in large corporations, and other professionals, many of whom detested the flower children and were made extremely unhappy by the possibility that their children might be—or were—having sex and using drugs. Again, the civil rights movement, while it included many whites, was dominated by blacks, both in the leadership and in the rank and file. Most of these people were laborers and could hardly be defined as hippies, although there were of course among them some who advocated the use of drugs, unlimited sex, and the ideals of communitarianism. Finally, causes like environmentalism and dietary reform have long histories which predated the hippie movement and lasted long afterwards: macrobiotic foodstuffs could be bought on New York City's fashionable Madison Avenue, hardly a haunt of the hippies.

It is true, of course, that perhaps the bulk of those who could be defined as hippies did favor desegregation, did hate the Vietnam war, did support the preservation of the environment, did eat all that wholegrain rice. Many of them marched, sang, demonstrated, and some of them had their skulls split by police billies at Columbia University, in the streets of Chicago, and elsewhere in 1968. But I am more concerned with another aspect of the hippie movement which has, it seemed to me, had more profound and longer lasting effects than the anti-war and civil rights efforts. This was the drive to build a new way of living, a counter lifestyle, which was supposed to act as an example for the rest of America and lead the nation into a new era of peace and happiness.

Defining who belongs to this group is not easy. There were millions of Americans at the time who put on beads and sandals on Saturday night and went out to smoke pot. There were many older people, attracted by an untrammelled lifestyle they felt they had missed out on in their own youths, who were drawn in, but not so far that they were ready to give up careers and professions. And there were many more who gave the hippie life a fling for a few days or weeks or months and then went back to their colleges or their jobs.

I think, however, that we can make a stab at isolating for study a hippie group if we look at those who actually joined the communes which were a phenomenon of the time.[53] These groups came in all shapes and sizes. Some were quite formally organized, others barely organized at all; some were drug free, others saw drugs as a central part of their lifestyle; some were rural, others based in cities.

But however much they differed, the members generally believed "that they were pioneering, carving out an environment where spontaneity, human intimacy, and love could replace the old values they were rejecting," says William H. Chafe in his study of the post-war period. And he adds, "By 1967, the new lifestyle had become a national sensation."[54]

But in fact, once again the attention given to the hippies by the press drastically exaggerated their numbers. It was estimated by George Hurd, a commune member, that in 1968 there were "hundreds" of such communes operating at that moment. They ranged in size from five to eighty people, but most of them, again according to Hurd, settled down at twenty or thirty. There was, however, always an enormous traffic of people who made brief stays, even briefer visits, or actually came as sightseers through the communes. Holiday had some three or four thousand people come for a week, a month, or more, just for the experience, over a period of about eighteen months. Testimony from other commune members indicates that this flow of part-timers was typical.[55]

If these figures are in any way correct, they suggest that over the years of the late 1960s and early 1970s, when the commune movement peaked, the number of people who committed themselves to the life for any length of time was less than 10,000. To this must be added an indeterminate number of short-timers having a brief fling with the hippie life, which certainly ran into the tens of thousands and possibly even the hundreds of thousands: in 1971, 36 percent of college students were interested in the "commune idea," according to one survey.[56] Finally, there was a mass of free-floating hippies, like the flower children who haunted the Haight-Ashbury area of San Francisco or the East Village of New York, who were living in unorganized pairs and groups in lofts and cold-water flats. How many there were in such places is impossible to know, but calculating the population density of these areas suggests that there could hardly have been as many as 25,000 hippies living even in a large enclave like the East Village. Leaving aside those who dropped in and out quickly, it is difficult to see how the hippie hard core could have run to as many as 500,000 people.

They were being drawn from an age cohort of somewhere around twenty-five to fifty million, depending upon where you want to draw lines. These figures are of course very rough, and can only be suggestive. What they indicate, however, is that the actual number of hippies during the peak years of the movement constituted only a small fraction of the young—not nearly 10 percent at the furthest stretch, and probably less than 5 percent.

What were the communes like? Some, like the Mike Metelica group in Massachusetts, were dedicated to abjuring most of the sins of the flesh. The members had given up drugs, alcohol, and even sex—only the dominating Metelica was having sex with the three or four women who belonged to the group. The group neither begged, bought, nor stole, but spent their days offering their labor free to local farmers who needed help, and counted on the Lord to provide, which He did, although not in any great abundance.[57]

But far more typically the communes operated on a laissez-faire principle with each member defining for himself what he would do. George Hurd said:

> It's probably the most difficult problem I'm faced with in terms of communal life—whether or not you define the trip. Most people are tired of rules and regulations. You hope that there'll be some kind of awareness so that they'll define their trips. I'm more in favor of an open commune, but define the trip. Usually it works if you have patience: they either come around or they leave.[58]

Similar views were held by people at other communes. A member of one in Buckland, Massachusetts, which had no organization, no leader, no work schedules, said, "No one wants a leader, and no one wants to be one." Another member of the same group said, "What the community has done for me is allowed me to treat other people as equals, not superiors or inferiors." And another: "We're trying to cut work to the minimal, so that people can sit there and write."[59] Again, Morning Star, in Mendocino, north of San Francisco, whose property was registered in the name of God, was "completely open for anyone to do anything," according to one observer.[60]

Not all the communes were so unstructured. The Hog Farm in the San Gabriel Mountains of California had a rotating work leader, and had rules against dope and would not take in runaways, both of which might have got them into trouble with the civil authorities.[61] But most groups were dedicated to the idea that everyone was to "do their thing," as the saying of the time was. The dream was that one person would take the children on nature walks, while another canned tomatoes, a third cut wood, and a fourth went into town and worked at a gas station to earn money for rice and macrobiotic grains. The reality was that the "thing" of a great many members of communes turned out to be drugs, sex, or lying in the sun. This was hardly true of all; there were plenty of serious, industrious people in communes who were willing to work hard to make the plan succeed. But there were numbers of others who were not. In particular, the communes tended to attract many who were having trouble making places for themselves in the mainstream of society—misfits, and a good many people suffering from one or another emotional disability.

Drug addicts in particular were a problem for many communes. For example, a lot of speed was used at Morning Star; there were knifings, and drug busts by local authorities as a consequence.[62] At the Free University of Montreal, according to member Paul Gregg, "The drug thing got to be a problem. Many of the people became heads. 'No man, let's not plan things. Don't take it seriously, if it happens it happens.' Then if you take it seriously they said, 'Don't push me.'"[63] In another instance, some of the faculty at Goddard College, a small experimental college then existing in Vermont, invited members of a commune to spend two weeks on campus in hopes that the contact would prove liberating for the students. Unfortunately the commune was drug-oriented, and instead of preaching the simple life, the members mooched off the student dining room, and destroyed property. College authorities wanted to remove them but worried about a potential bad reaction from the students. The commune stayed on until one member went after another with a pistol in an argument over a woman; after that the group left.[64]

Drugs contributed to a second problem many communes faced: getting everybody to do a fair share of the work. George Hurd, reporting on the California group called Holiday, said, "There were maybe a half dozen people who always saw what had to be done, but most of the others did a share of work. A few goofed off completely."[65] Such was frequently the case elsewhere.

A third matter that was a constant thorn was sex. There was a tendency, as has always been the case among humans, for people in the communes to pair off. This tendency, however, was limited by the simple fact that males almost invariably outnumbered females—according to George Hurd, typically by two to one. Most people in communes recognized that this male-female imbalance was a potential bomb primed to explode. In many instances unattached males attempted to contain their sexual feelings for the general good, but it did not always work. At the Free University of Montreal, Paul Gregg said, people tried to be careful about sex for fear of jeopardizing the commune, but "When the holding back broke down, then there was trouble."[66]

In fact, the communes were facing precisely the problems that over thousands of years human culture was developed to solve: distribution of power, assignment of sexual rights, and division of labor. But they were committed, most of them, to the structureless society in which nobody told anybody what to do. The consequence was that any kind of behavior had to be tolerated, no matter how much it disturbed others, or even threatened the existence of the commune. Dennis Lee of the Rochdale College in Toronto said that about 10 percent of the members were "acid heads" or used other drugs. He said, "They're very welcome, but if Rochdale turned into that, I'd be sad."[67] At the Pennsylvania commune, George Hurd said, "We have one guy who digs guns, and every time a gun goes off there's a hassle because the rest don't like guns. A lot of people want to avoid the hassle. We don't go for rules."[68] It was the same with sex: whatever people may have felt, their code required that they accept infidelity, mate-swapping, promiscuity, even when this kind of behavior was causing considerable friction in the group.

In practice, of course, leaders did emerge, and in the most effective communes social pressure, however much it would be denied, was brought to bear on members who were disruptive or not pulling their weight. Nonetheless, an astonishing number of the communards were genuinely committed to the ideal of the egalitarian, unstructured society, and tried hard to make it work. In doing so they were taking the ethic of self, which had been deepening its hold on the culture through the century, and making it not merely *an* ideal, but *the* ideal. Even when the very existence of the group was threatened, there could be no compromise with the needs of the selves of the members. The self was to come first, last, and always.

This was an extraordinarily rare, perhaps unique, ideal on which to base a social system. Everywhere, not merely in human societies, but in most other primate groups as well, the integrity of the group is a primary concern. Tolerance for deviance from the norms varies enormously from group to group; but a line is always drawn when the solidarity of the group is at risk. I do not believe that there has ever existed a successful human society in which the needs of the individual members were invariably put ahead of those of the group. Indeed, one of the primary functions of a culture is to adjudicate between the needs of the group and those of the individuals who make it up.

But in some of the communes the idea that the self came first was not merely understood, but explicit. Members of the Hog Farm saw their commune as having no past and no future: they were prepared to let it disintegrate when "it no longer feels good."[69] And it is therefore not surprising that few of them lasted very long.

In the end, this contradiction at the heart of the commune ideal killed it. The *group*, which needed wood cut and corn hoed and shoes mended and children taught, could not be reconciled with *selves* who needed to write novels or play frisbee or smoke dope or lie in the sun. This contradiction was obvious and should have been apparent to the people involved; but it was not, because as a whole the movement lacked intellectual rigor. Indeed, it was almost purposefully anti-intellectual. As the famous Port Huron declaration put it, "We regard men as infinitely precious and possessed of unfulfilled capacities for reason, freedom and love."[70] This was a nice thing to believe; but the entire history of humankind, to say nothing of modern psychology, anthropology, and sociology would not lead one to that conclusion. These people had chosen to ignore history, even the history of their own ideals. It does not appear, for example, that many of them studied the well-known utopian communities of the nineteenth century, such as Brook Farm and the Oneida Community, to see what could be learned from them. Operating on faith, they bulled ahead; but they were trapped by the competing needs of self and community. As a consequence the legacy of the hippies was not a new social program, but the popularization of drugs and a wave of teenage promiscuity.

This, I recognize, is a very harsh judgment to make of these people. They did see, perhaps more than most Americans, that there were serious wrongs in their society. William Chafe says, "Rejecting the passivity of their countrymen in the face of injustice, they pledged to revitalize American democracy, to create communities where work could be meaningful and leisure fulfilling, and to fight for an America where corporations would become more publicly responsible, politics more representative of people's needs."[71] In this they were more sensitive than most Americans and certainly more idealistic. And of course many of them took the very real risks associated with demonstrating, evading the draft, and such. But their desire for the unlimited self made effective group action impossible. And they foundered, and went down.

But their lifestyle and the philosophy it grew out of remained alive in the minds of Americans. Unlike the Beats, the hippies proselytized for their ideas, at least to the extent of advocating their ways as *exempla* for the rest of the world. And it seemed to many people observing them that their ideas and their lifestyle had social virtue. Perhaps these young people were onto something. Not everybody over thirty approved by any manner of means: most communes faced a good deal of hostility from local communities, and the majority of Americans did not accept the drug culture. But there was enough interest in the press to make the movement and its philosophy widely known. For example, in the five years from 1967 through 1972, the *Reader's Guide to Periodical Literature* lists 163 articles about "hippies." By comparison, in a similar five-year period ten years earlier, the *Reader's Guide* carried only thirty-five pieces on "beatniks." Furthermore, the Beats were treated as a literary movement, and pieces on them in the newsmagazines, for example, were generally carried in the book section; pieces on the hippies were usually carried as general news or features and would get a much larger readership.

It was, however, not just the magazines. The so-called counterculture provided good "visuals" and received a lot of television coverage. And beyond the news media, there was the outpouring of rock music, which was seen as part of the movement and acted as a kind of spokesman for the hippie idea, with its constant references to the virtues of drugs, sex, and dropping out. By 1970 it was hardly possible to be an American and not be aware of the counterculture and what it stood for. And it would have been impossible to expect that its ideas would not be picked up by millions of young Americans, who found in them not only social virtue, but a lot of fun as well.

CHAPTER 16

The End of Morality

The Beats took the new philosophy of the self, which had budded in the second decade of the century and blossomed in the third, and pushed it to an extreme, actually trying to live as if it were possible to ignore the needs of any but the self. The hippies, next, both through the formal organizations of the communes and the looser, informal groups that gathered in the East Village, Haight-Ashbury, and elsewhere tried to put together social organizations built on the self. What had originated in the early part of the century as a personal philosophy was to be tried as a political system; not only were the hippies organizing social groups around the ethics of the self, but many of them hoped that it could be applied to grave national and international problems.

The extreme deviance of the Beats and the unconventional lifestyle of the hippies would appear to contradict my earlier statement that the years from 1945 to the early 1970s—what I have been calling the postwar period—were a stable time in America which produced no real revolutions.

We must remember that neither the Beats nor the hippies were America: the Beats constituted a minuscule fraction of the population, and the hippies, although more numerous, did not amount to more than a small percentage even of the generation they were supposed to represent. Both movements were made to seem more significant than their numbers suggested by the media, which found them good copy.

But while the public was hearing a great deal about the colorful new groups and their ideas, it was not, in the period we are speaking of, taking them up. For when we look at the figures for sex, drugs, and drink, the commodities which were central to the thinking of the deviant groups, we see that they remained remarkably stable though this whole period of nearly thirty years.

We can begin with sex.

One great advantage of the Kinsey Reports was that they stood alone, a World Trade Center rising out of a broad plain. With all of their faults, the Kinsey volumes are the only reference points we have for the period they covered, and we are forced to use them, however gingerly. But the deficiencies of the Kinsey Reports inspired a number of successors hoping to correct what were seen as biases and skews in the work. Some of them attempted to survey what are known as "probability samples"—that is to say, groups of people chosen to be representative of the population as a whole, so that the results can be projected to the society at large. Others tried to get away from what many people viewed as the rather mechanical approach to sex they found in the Kinsey Reports—which measured discrete behaviors, such as orgasm, masturbation, and the like—the idea being to take into account less tangible matters like attitudes and feelings.

There was, however, no consistent agreement as to how studies should be done, and the consequence was that the spate of sex studies that appeared during the post-war period are not readily comparable, either with the Kinsey Reports or among themselves. We are confronted not merely with apples and oranges, but with pineapples, grapefruit, and persimmons as well. Moreover, many of these studies used limited and atypical populations, most especially the college students that social scientists had ready to hand. It is very difficult, thus, to compare Winston Ehrmann's 1947–50 study[1] of some 1,000 unmarried students enrolled in Marriage and the Family courses at a southern university with, say, Lester A. Kirkendall's 1950s' study[2] of sexually experienced males at a college some 3,000 miles away. Nor is it possible to project these studies across the population as a whole: the samples used are unrepresentative. These studies, and others like them, are valuable in different ways and have things to teach us. But because they were designed to examine somewhat different things, we cannot use them to track changes in sexual habits and attitudes.

Among them, however, are two studies which can help us to make these temporal comparisons. One was a study made by Ira L. Reiss,[3] considered by many authorities to have done some of the most important work in sexology in the period, which used a probability sample—that is, a demographically representative group—in two stages in 1959 and 1963. The second is a Kinsey Institute study, made in 1970 by a team that included Albert D. Klassen, Colin J. Williams, and Eugene E. Levitt,[4] which again used a probability sample. (The report was not published until 1989, because of contentions among the authors over division of credit and other matters.)

Whatever their limitations, both studies are based on sound statistical sampling methods and can therefore be generalized to the population as a whole. And both say flatly that there was no revolution in sexual behavior in the post-war period, and only a relatively modest one in changes of attitude. To take the chronologically earlier of these studies first, using samples taken in 1959 and 1963, Reiss begins by saying what we have earlier observed, "Many studies support the contention that an increase in sexual behavior of several types occurred during the 1920s."[5] He goes on to say, "Kinsey showed that little change had occurred in coital behavior since the 1920s, whereas attitudes have changed more since then"; but, "There is consistent evidence from past studies, such as those by Kinsey and Terman, to show trends toward increased permissiveness at all levels of sexual behavior"[6]—permissiveness being a measure of attitude more than of behavior.

This is, of course, what we have seen—a change in attitude dating back to about the time of World War I, with a change in behavior following about ten years later, and then slow upward movement in the direction of permissiveness, with somewhat higher rates of premarital sex, extramarital sex, petting, and declining ages for sexual experimentation.

These same tendencies continued through the decades immediately following World War II. Reiss says, "The parents who were courting in the late 1940s were probably quite similar to their children courting in the late 1960s. The major studies show little evidence of behavioral changes in coital rates, but they do show changes in petting rates. Basic changes in the last forty or fifty years have more likely been in the area of attitudes than in behavior. Americans are more open in their discussion of sex, and permissiveness is more respectable today."[7]

Most interesting, Reiss found in his subjects, studied in 1959 and 1963, a very considerable congruence between attitude and behavior. For example, about two-thirds of the college women who believed that they should go no further than kissing did not go further; 78 percent of those who put the limit at petting actually adhered to the limit. The congruence was not perfect: 31 percent of those women who did not accept coitus actually engaged in it, and 36 percent of those who did accept it did not engage in it.[8] But those who deviated from their own standards did so primarily because they happened to be in love, or not in love, at the appropriate moment. One of Reiss's important conclusions—which other observers had noted—was a new acceptance of "permissiveness with affection"—that is to say, a lot of things were more acceptable for people who were bonded in one way or another than they would be for casual acquaintances. Reiss concluded for the period under discussion that behavior and attitudes were "closely intertwined and are predictive of each other even though there is not a perfect one-to-one relationship."[9]

At the time the Reiss study was made, although it was already clear that a new openness about sex was at work in America, nobody was talking about a sexual revolution. But the Kinsey Institute study by Klassen et al. was based on a survey conducted in 1970, precisely at the moment when the supposed sex revolution should have been in triumphant possession of the land. It was, like the Reiss survey, done on a carefully drawn demographic sample of the American population, and its findings can therefore be generalized across the culture as a whole. And the authors flatly conclude, "Our data have shown that patterns of sexual morality in the United States in 1970 tended to be quite conservative. The findings do not support the contention that a 'sexual revolution' had occurred in 1970 or is now occurring in the United States."[10] And they add, "Traditional structures of meaning have remained powerful and continue to provide the social context of sexuality for most Americans."[11]

Unfortunately, the authors of this study lump any kind of heterosexual behavior leading to orgasm on the part of at least one of the partners, which does not allow us to make direct comparisons of behavior with other studies, but the figures nonetheless support the authors' contention that we were not seeing in 1970 any dramatic liberalization of sexuality. Sixty-six percent of women in the study had never masturbated;[12] 60 percent had not had a heterosexual experience before marriage leading to orgasm even by the partner,[13] and for 60 percent of those who had had such an experience it was with only one partner, usually the man they would eventually marry.[14] Or, to turn the figures around, only 25 percent of women at the time had any very considerable sexual experience before they married. Males, as always, had had more premarital sexual experience: about 80 percent had had at least one heterosexual experience leading to orgasm before they were married;[15] but, again, only about a third of them had had this experience "fairly often," and over 40 percent of the males had had only four or fewer sex partners before they married.[16]

Older people of the time were consistently less approving of premarital sex than younger ones, by significant margins, but what is most surprising is how many of the post-war generation disapproved of it: over a third of the post-war generation felt that even an adult male who was in love should not engage in sex before marriage, and over half disapproved of premarital sex for a teenage girl who was not in love with the male at the time. Almost two-thirds of this younger generation disapproved of extramarital sex; and about three-quarters disapproved of homosexuality under most circumstances. Some 40 percent of this post-war group disapproved even of masturbation. Not surprisingly, the women in the sample (taken from all generations) favored tougher strictures than the males. For example, over 90 percent of the women surveyed disapproved of premarital sex by a teenage girl who was not in love, and almost three-quarters felt that it was wrong when she *was* in love. Indeed, over 60 percent of American women in 1970 felt that sex before marriage was wrong even for an adult woman who was in love.[17]

Taking the Klassen, Reiss, and other studies altogether, it is abundantly clear that no dramatic revolution in sexuality took place in the 1960s. The fact that some half of Americans still believed that premarital sex was wrong under virtually any circumstances as late as 1970 and larger percentages disapproved of it for people who were not in love, or who were young, suggests that the sexual morality of America at the time was still essentially conservative. The authors of the Klassen study said at the time, "Except for masturbation and for premarital sex between people who are in love, our data suggest that a majority of Americans are 'moral absolutists' in that they see these behaviors as *always* wrong."[18] And Daniel Yankelovich, speaking of the very late 1960s, said that the new sex code "is confined to a minority of college students."[19]

Nonetheless, the Klassen figures do show increasing permissiveness in younger groups, both in attitude and behavior. Overall, what we are seeing once again is the continuation of the long-term trend, which reaches back to the end of the Victorian era in the first years of the twentieth century, toward a gradually growing liberality in sexual behavior.

A far more significant revolution of the post-war period was the swift, and astonishing acceptance of the use of drugs. Drugs were hardly invented by the Beats. They have a long history in mankind, and certainly there was no shortage of drugs in the United States in the nineteenth century. Most of the nostrums sold to millions of women for "female complaints" in the Victorian age were laced with opium or cocaine derivatives, and drugs were widely available in the vice districts and elsewhere. The Philadelphia Vice Commission reported in 1910 that 68,000 pounds of cocaine were consumed in the United States annually by 150,000 Americans and 120,000 resident Chinese.[20] Much of this drug consumption was legal; the Chicago City Directory for 1911 carried a prominent ad promising "Drug Habit Cured in 10 to 14 Days,"[21] suggesting that drug addiction existed among the middle-class people who would be likely to use a city directory. According to one authority, Victorian drug use fell in the 1890s, probably due to the economic crunch, and then rose through the 1900s until "America went on to become the world leader in narcotics."[22]

But before the 1960s drug use was confined mainly to the vice districts, the underworld, the jazz world and show business generally, and certain immigrant groups, like the Chinese, who had a tradition of tolerance for drug usage—leaving aside, of course, the women who were drug-users unbeknownst. Even during the time when the Beats were rising into prominence in the 1950s, few ordinary Americans had ever even *seen* a drug, much less used one—and that includes marijuana. Drugs, for most Americans before the 1960s, were a distant, dangerous, even evil phenomenon associated with murderous tong gangs, madmen, and the truly depraved. A survey of New York City made in about 1970 showed that while 20 percent of the twenty-one to twenty-four-year-old cohorts had used marijuana (all of which use must have come in the 1960s), only 4 percent of those ten years older had done so, and only 1 percent of those thirty-five or older.[23] We can assume that much of the experience of drugs that these older people had had came in the 1960s, which suggests that, if this study is at all reliable, only minuscule percentages of Americans—probably fewer than one of a hundred—had used drugs at all before the 1960s. Americans born before World War II simply did not use drugs, at least not until the 1970s.

The drug boom was produced by the post-war generation. Unfortunately, we lack good figures for the extent of drug use in America for this period, for it was not recognized as a problem immediately, and it was not really until the 1960s and especially the early 1970s that social scientists began seriously to measure drug use. However, a Bureau of Narcotics and Dangerous Drugs report showed that between 1960 and 1969 there was a "51 percent increase in the number of active addicts and the number of addicts under 21 years of age increased by 173 percent." The National Institute for Mental Health estimated that by 1970 there were 100,000 to 200,000 drug addicts in the United States.[24]

These figures are for hard drugs and do not include regular smokers of marijuana. However, extrapolating from the figures produced by the New York survey, we can estimate roughly that use of marijuana increased somewhat during the late 1950s, doubled during the first half of the 1960s, and doubled again by 1970. By that year another study estimated that between twelve and twenty million Americans had used marijuana at least once in their lives;[25] and since by that year only tiny numbers of older Americans had used the drug, most of those twelve-to-twenty million American pot-smokers were in the younger cohorts, a notion borne out by a good survey which said that by 1968 some 45 percent of American college students were opposed to the legal limits on marijuana use.[26]

This sudden acceptance of drugs, which took place in a span of about ten years, is one of the most dramatic cultural shifts to occur during the American twentieth century. A behavior pattern that had been universally anathematized throughout the entire society, except in a tiny fringe under-class, overnight became accepted and widely practiced in a new generation, and was even beginning to spread into older cohorts as well. But the astonishing growth of the drug culture in the 1960s should not be allowed to mask the fact that it remained, for the moment, the practice of a minority, the largest proportion of them, prob-

ably, college students and recent college graduates (or college drop-
outs). A large majority of Americans opposed the use of drugs by
anybody, and the institutions of the society continued to bear down on
it. Religious leaders inveighed against drugs; federal, state, and local
governments arrested not only drug sellers, but users; and schools and
colleges imposed sanctions on students caught using drugs. There were,
of course, those in what was seen as a "counter-culture" who propagan-
dized for open drug use, among them the leaders of the hippie movement and *Playboy* publisher Hugh Hefner; but these were a distinct
minority. As was the case with the new sexual permissiveness, in the
1960s drugs were something more frequently read about than actually
used.

One justification for the use of drugs, especially marijuana, which
was then seen as harmless, was that it reduced the consumption of alcohol. That, however, was not the case. It is probably worth at this
point reviewing American drinking habits over the long period we have
been examining. According to the generally accepted figures published
by Lender and Martin, in the eighteenth century Americans had been
drinking around six gallons of absolute alcohol per person of drinking
age, figured to be fifteen years of age or older. This figure rose rapidly
through the early decades of the nineteenth century as an increasingly
isolated population spreading through the frontiers escaped social controls; by 1830 the per drinker intake of absolute alcohol was 7.10, over
20 percent higher than it had been forty years before.[27]

Then, as the temperance movement of the new Victorian age took
hold, consumption of alcohol dropped precipitously, falling by more
than half between 1830 and 1840, and continuing to drop decade by
decade until by the 1870s it was down to 1.72 gallons per drinker, less
than a quarter of the peak a half-century before. Thereafter drinking
rates began to rise, in the main due to the arrival of immigrant groups,
some of whom had traditions of fairly heavy drinking. The anti-Victorian spirit which was sweeping the land in the early years of the century
contributed to rising drinking rates in the native population, especially
in the burgeoning big cities, to a peak of about 2.60 gallons per drinking age person in the years just before World War I.

But countervailing pressures of the Prohibition movement, which was
drying up state after state, forced a reduction in drinking; and national
Prohibition pushed drinking levels even lower, probably to less than a
gallon of pure alcohol per capita—possibly the lowest drinking rate to
ever obtain in an industrial society. With the ending of Prohibition,
drinking rates began to rise again, leveling off in the post-war years to
about two gallons of absolute alcohol per drinker, a rate that remained
remarkably consistent from the mid-1940s into the mid-1960s. And at
this point drinking rates began to rise sharply. The steady pattern which
had continued from, roughly, 1942 to about 1962 was broken; in the

first half of the 1960s drinking rates rose by a relatively insignificant 7 percent. But in the second half of the decade the increase was 13 percent.

The substantial increase in alcohol intake which came at the end of the 1960s was certainly, in part, a result of the great prosperity of the period. Half of the jump was due to a 28 percent increase in the consumption of hard liquor; and although both wine and beer intake showed increases, it is fair to assume that, with more money in their pockets, Americans were, temporarily, moving back toward the hard liquor era of 200 years earlier: 1971 was the only year since 1890 in which Americans got more of their alcohol from hard liquor than from beer.

When we look at what I have been calling the post-war period—the years from 1945 to the early 1970s—we see a remarkable consistency in the country's basic lifestyle. Throughout most of this twenty-five-year period Americans each year drank almost exactly as much as they had the year before; the sexual behavior and courtship patterns of the young proved to be very similar to those of their parents; and although in the later years of the period drugs were moving out of the underworld into the mainstream, they were by no means acceptable to the vast majority of Americans. There had been on all fronts a continuation of the gradual liberalization of attitude toward the sensory pleasures, with perhaps a somewhat sharper rise in the very last years of the period. But there was no revolution: the kinds of things that Americans did in 1965 were very similar to those they had done in 1945. The most significant indicators of this stability are the rates of marriage and divorce in numbers per thousand of population: at the beginning of the period both were relatively high, reflecting the readjustments people were making after the dislocations of World War II, which occasioned a number of spur-of-the-moment marriages and quick divorces. But over the period as a whole, divorce and marriage rates held very steady; from 1950 to 1965 the marriage rate fell 15 percent, basically reflecting the spurt in family formation that took place after the war; and in the same period the divorce rate *fell* by 4 percent.[28]

While it is true that the last years of the 1960s saw the beginnings of new tendencies in American life, the real revolution in *behavior* took place in the 1970s, and especially in the early years of the decade. Let us look at the figures. To start with, between 1970 and 1980 the percentage of our gross national product that went for amusement and recreation increased by 40 percent.[29] A lot of this money was spent on television sets: in the peak years of 1972–73, when the revolution in behavior was surging forward, Americans bought some 34 million television sets,[30] even though 97 percent of American homes already had a set. The amount of time people were spending in front of their sets startled even sociologists. In the mid-1970s yet another team of investigators returned to Muncie, Indiana, to produce another follow-up to the original *Middletown* study. They found that in 1976 Middletown families were watching television 28 hours a week. "No large population anywhere had ever spent so much of their time being entertained. We could try to explain away these staggering statistics by supposing that people keep their sets turned on without actually watching them, but viewers in Middletown, at any rate, are able to recall the programs they claim to have watched, even though most of them have developed the trick of doing housework or homework at the same time, and intermittent family conversation accompanies television viewing more often than not."[31]

Nor was it just television: in Middletown movie attendance dropped by about 75 percent between 1950 and 1970 as television occupied more and more of people's time. In about 1970 the trend began to reverse itself: "By 1977, Middletown had a larger array of movie theaters than ever before," among them a pornographic theater.[32] The same was true elsewhere: movie box-office receipts (discounting for inflation) rose by 20 percent in the 1970s.[33] If this were not enough, in Middletown magazine circulation was up; attendance at basketball games, Indiana's primary sport, was up, participation in sports "showed spectacular growth."[34] In sum, the amount Americans spent on "amusement and recreation services," once again discounting for inflation, rose by almost 50 percent in the 1970s.[35]

Given this, it is difficult to see how people found time to drink more than they had, but they did. According to the Lender and Martin figures, between 1970 and 1978 consumption of pure alcohol rose by 17 percent. Again, this was despite a slight decline in the rate of hard liquor drinking from the peak year of 1971.[36] From 1970 to 1980 per capita consumption of beer increased by 31 percent, wine by 61 percent.[37] Actually, after the initial jump in alcohol consumption in the years around 1970, rates held fairly steady until the last years of the 1970s, when they began to climb again to yet another peak.

But the most spectacular growth industry of the 1970s was drugs. One study says that between 1972 and 1977 the number of Americans in the eighteen- to twenty-five-year-old group who had used cocaine more than doubled, from 9 percent to 19 percent.[38] According to this same study, the number of eighth graders who had smoked marijuana again nearly doubled in the two years between 1970 and 1972. The number of high school seniors who smoked marijuana daily increased from 6 percent to 11.[39]

Another study reported that between 1972 and 1974 there was an increase in every type of drug use in the twelve- to seventeen-year-old group.[40] The numbers who used cocaine and marijuana doubled in that short period. Increases in the number of people who were using drugs at older ages were less dramatic, but they continued to climb through the early and mid-1970s. In 1975, 54 percent of high school seniors had used an illicit drug at least once; by the next year the number was 58 percent. Nor was drug use a rare, once-in-a-lifetime experiment for these students. In 1976, 35 percent of them had used an illicit drug within 30 days of the survey, 32.2 percent had smoked pot in the same period, up from 27.3 percent the previous year.[41]

Perhaps the best study of drug use among high school students is an ongoing survey being conducted for the National Institute on Drug Abuse by the University of Michigan Institute for Social Research. It found that "marijuana use showed a dramatic rise during the 1960s and the early 1970s. Our own data indicate that the trend continued strong from 1975 through 1978."[42] Their data show that between 1975 and 1979 the number of high school students who used cocaine rose from 5.6 to 12 percent; and that by 1979, 2 percent were using cocaine on a weekly basis, with a tenth of that number using it on a daily basis.[43] And these figures, the authors say, are biased downward, because some 15 to 20 percent of the young people in their schools had dropped out of high school by the time of the survey, and so were not included in it.[44] These drop-outs, they assume, were likely to be heavier than average drug users. But marijuana was the drug of choice: by 1979 60 percent of high school seniors had used the drug at least once.[45]

As with drugs and alcohol in the Roaring Seventies, so it went with sex. As we have seen, in the late 1960s there was a continuation of the gradual, long-term trend toward liberalization of sexuality, with attitudes generally out-pacing behavior. For example, a study of a group of college students by Ira E. Robinson and Davor Jedlicka showed no change in the percentage of males who had premarital sexual intercourse between 1965 and 1970 surveys, and a relatively modest rise in the number of women who had done so; but the change in *beliefs* about what was right and wrong in regard to sex had showed a considerable change in the direction of permissiveness. The number of these college students who believed that premarital sexual intercourse was "immoral" dropped by *half* between 1965 and 1970, and the numbers who felt that people who had sex with "a great many" partners were immoral dropped by about a third over the same period. This was indeed a dramatic shift in attitude to have occurred in so brief a period; but the changes in behavior created by the shift in attitudes did not show up until later. Between 1965 and 1970 the number of males reporting that they had premarital intercourse actually dropped fractionally; in the next five years the rate increased by 14 percent, and continued to rise at a somewhat slower pace through the decade. Similarly, the big jump in pre-marital intercourse for the female students came in the early 1970s, when the numbers who were involved jumped by 54 percent, once again a dramatic increase in a behavior pattern for so brief a period. And as with the men, throughout the decade the numbers of young women in this sample who engaged in premarital sex continued to rise.[46]

Other studies of college students show the same trend. One 1974 study indicated that 29 percent of the males of a group of college students and 18 percent of the females had lived with people of the opposite sex and that 71 percent of the males and 43 percent of the females said they would do so if the occasion arose.[47] At another college the figures for those who had lived together were 33 percent of the males, 32 percent of the females. Say the researchers, "When the opportunity is available, the peer group is supportive, and if the couple wants to live together, they probably will."[48] And this was a nationwide trend. By the mid-1970s one study of college students reported that about a third of the males and a quarter of the females actually had lived with a sexual partner, and that 40 percent of those who had not done so would consider it.[49] What had been a rare, even scandalous lifestyle in 1960 was by the 1970s commonplace.

It is particularly important for us to realize that the new attitudes and new behaviors which were appearing among college students were rapidly being taken up by non-college youth in the 1970s. Daniel Yankelovich, a psychologist specializing in studies of American attitudes, says, in a major study of American youth of the 1970s, "The new sexual morality spreads both to mainstream college youth and also to mainstream working class youth."[50] His figures show that in 1969, 34 percent of college students thought that casual sex was wrong, while among working-class youth the figure was 57 percent. By 1973 the percentage of working class young who believed that casual sex was wrong had dropped to 34 percent, exactly the figure for college students four years earlier.[51]

Furthermore, it was seeping down to the brothers and sisters of the college students who were still in high school. A study of high school students in two representative communities showed an increase in coitus between 1970 and 1973 for high school boys from 27.8 percent to 33.4 percent, and for girls from 16.1 to 22.4 percent, again an astonishing behavioral shift to have taken place in the space of four years. The numbers of girls who had had more than one sexual partner in the high school years went from 5.7 percent in 1970 to 8.9 percent in 1973. For boys the figures rose from 14.2 to 19.4 percent. The researchers say, "There was an increasing likelihood that girls would experience intercourse at earlier ages. Our data indicated that this generalization is also applicable to boys."[52] By 1973 an astonishing 28 percent of boys had had sexual intercourse at thirteen or younger; for girls it was 10 percent. By age sixteen about half of both sexes were involved in heavy petting, and between 1970 and 1973 the numbers of boys who had intercourse by age sixteen had risen from 31 to 38 percent, the number of girls from 23 to 31 percent. By the mid-1970s over a third of female high school seniors had had sexual intercourse. Summing up, one group of researchers concluded, "The evidence suggests that the premarital sexual revolution which was reported to have begun during the late 1960s has actually accelerated during the 1970s."[53]

One thing most observers noted was that sex, drugs, and alcohol were intertwined. The students, both in high school and college, who had had sex were also likely to be the ones who were drinking and using drugs. Arthur M. Vener and Cyrus S. Stewart, in one study, report that "high correlations were found between sexuality and the use of illicit drugs."[54] Other studies concur. The point is not that drinking and using drugs triggered sexual behavior, but rather that a certain mind-set in the young inclined them toward a general permissiveness. Vener and Stewart conclude, "The findings underscore our initial impression that substantial change has occurred in the social climate of the school system" between 1970 and 1973.[55]

In fact, the changes were running through the entire society. A 1976 survey indicated that 32 percent of middle-class American women in the thirty-five- to forty-year age bracket had had extramarital sex at least once, and, even more surprising, that 56 percent of these women *approved* of the idea of extramarital sex, at least for themselves.[56] "American women who have experienced EMS [extramarital sex] tend to perceive the opportunities for EMS as readily available . . . They are inclined to feel that the disclosure of EMS will not incur severe sanctions . . . such as divorce, loss of job, or damage to their reputations. The prevalence of premarital sexual experience . . . and exposure to pornography . . . in this group of women indicates that they have availed themselves of opportunities for nonconventional sexual activities . . . They show freedom from traditional familism . . . and above all, positive attitudes toward EMS."[57] The author of this study clearly believes that a willingness to commit adultery on the part of her subjects is liberating, and she suggests that the women in her study have the same attitude. The subjects of this study were born in the late 1930s, and were marrying in the mid-1950s and after. Earlier studies of this cohort show them to be relatively conservative, and it can be inferred that by the mid-1970s they were being affected by the new morality which was springing up all around them. Further evidence that the change in attitude was moving through the society comes from a study of 1954 college graduates and their daughters. These mothers belonged to the same cohort of women in the adultery study and showed the conservatism toward sex prevalent at the time: 94 percent of them were virgins at their high school graduations, and almost half of them had not petted more than occasionally.[58] (These were college graduates, who in that period characteristically were slower than other groups to engage in sexual activity.) But 55 percent said that their daughters had helped to liberalize their attitudes toward sex.[59]

Studies like the foregoing indicate that attitudes toward sex were changing dramatically during the 1970s. The idea of "permissiveness with affection," which seemed a major switch in sexual attitudes in the 1960s, was now being abandoned in favor of a widespread acceptance of what has been called by some "recreational sex," that is to say, casual promiscuity. For large numbers of Americans there were no longer any rules in regard to sex.

How much things had changed is made clear by the case of Ralph Ginzburg and *Eros* magazine.[60] Ginzburg, operating on the *Esquire-Playboy* principle that you could get away with a good deal if your publication was sufficiently arty, printed *Eros* on slick paper and bound it in hard covers. Unfortunately, he included in one issue a colored photograph of a naked man and a woman, from about knee to shoulders, in a close embrace. The fact that the man was black, the woman white, was not irrelevant. From the position it was obvious that the couple could not be copulating, but despite this Ginzburg was prosecuted by the United States government, and went to jail.

By 1970, only a few years after Ginzburg was jailed, collections of erotic art, far franker than anything Ginzburg ever printed, were being openly published; and very quickly thereafter a booming business in pornography developed throughout the United States, with so-called "adult" book shops openly selling material which would have got the proprietors arrested not many years before. In 1971 *The Sensuous Man* and *Any Woman Can!*[61] were on the *New York Times* best-seller list, and were joined in 1973 by the Comforts' *The Joy of Sex*,[62] which was filled with pictures of sexual positions. In 1976, Shere Hite published *The Hite Report,* featuring endless and detailed descriptions of how women masturbated;[63] the book was given serious reviews in major media. Ten years earlier the book could not have been published—certainly not by Macmillan—and would hardly have been reviewed by the *New York Times.*

Although the question has not been thoroughly studied, there is little doubt that the new sexual freedom being advanced by the hippie generation and put into practice by their younger siblings sent a shock wave through the society. The older generation had by and large accepted the more conservative morality which permitted sex only in the context of a relationship, and then preferably when there was the expectation of marriage. Probably the majority of older women and a considerable percentage of older men had never had sex with anybody but their spouses; and suddenly, as they found themselves in their late thirties, forties, and fifties, they discovered that the old rules no longer applied. Many of them felt that somehow they had been cheated, and quite a few of them set out to rectify matters.

The inevitable result was a sharp upward movement in the divorce rate. Stable for a generation, from 1945 to 1965, it leapt 40 percent in the last years of the 1960s, another 37 percent in the years from 1970 to 1975, and continued to climb through the rest of the decade, until by the end of the 1970s the divorce rate was double what it had been through the post-war years of the 1950s and 1960s. Through the critical last years of the 1960s and the first years of the 1970s, the figures for divorce climbed 8 percent annually, once again a dramatic shift in the culture.[64]

The change in American sexual morality of the 1970s surpassed in kind and extent anything that had gone before. The sex revolution of the 1920s was, by comparison, about as bold as a church social. But we must bear in mind that the great shift in sexual morality was merely part of a massive change in the behavior of American society as a whole. Underlying the sudden, dramatic arrival of permissiveness toward sex, drugs, drink, and the like was the collapse of the last defenses against the cult of the self: by the 1970s if it was not good for "number one," it was not good. It was no accident that one of the most successful new magazines of the time was deliberately called *Self* by a publisher who knew precisely what he was doing. Or, as Daniel Yankelovich has put it, in the 1970s the young were "developing a new agenda of social rights. By social rights we mean the psychological process whereby a person's wants and desires become converted into a set of presumed rights."[65]

That was precisely the point. What had begun as a struggle to free the self from the constrictions of the Victorian code had evolved into a new social code, a conscious philosophy with the self at the heart of it. Indeed, it had gone even further than Daniel Yankelovich suggests: By the 1970s many Americans had come to believe that they had not only a right to put themselves first, but that they had a positive duty to do so. People now felt that if they did not feel good, were not getting "the most out of life," were not expressing themselves through satisfying relationships, fulfilling jobs, exciting avocations, there was something wrong with them.

CHAPTER 17

The Institutionalization
of Selfishness

We have seen, through the course of this book, a steadily rising curve of self-indulgence through most of the 20th century, on almost any measure you can name. Until the 1970s, however, the movement was steady and gradual—a little more sex outside of marriage, a little more time spent being entertained, a little more drinking, and the rest of it, every year. The ethic of the self crept into American culture a step at a time. Why, then, the sudden, sharp upswing in self-indulgence which was manifest in the culture by 1973?

It is difficult to analyze such things, but there were at least five factors operating which, it seems to me, may have worked together to produce this effect. One of these was the astonishing and unprecedented prosperity of the country during the post-war years. Using constant 1982 dollars to eliminate the effects of inflation and the deflation of the Depression period, we find that the "disposable personal income" per capita in 1929 was $4,091 (in 1982 dollars). The figure fell through the bad years of the Depression, but by the end of World War II in 1945, it had climbed up to $5,285. It actually fell slightly through to 1950 as adjustment to peace caused some dislocation in the economy; and then it took off: $6,036 in 1960, $8,134 in 1970, $9,722 in 1980.[1] That is to say, in 1980 the average American had twice as much *real income* as his parents had had at the end of World War II.

The young people who were coming into the work force in the early 1970s, and thus able to make decisions about how they would spend their time and money, had been born, roughly, between the end of the war and 1955. They had been raised in a prosperity no society anywhere in the history of the world had ever provided for its general populace. As children they lived in the new houses that were mushrooming in the suburbs around the great cities, chose from an array of food and drink that would have excited Henry VIII, were driven hither and thither in new cars complete with radios and air conditioners, were supplied with closets full of toys and bureaus filled with new clothes which they outgrew before they wore out. They took it for granted that they should have dancing lessons, guitars, their own radios, record players, even television sets. Instead of playing baseball on corner lots or in cow pastures with scuffed baseballs stuffed with newspapers, they played on well-groomed fields, wearing handsome uniforms and using equipment superior to that available to major league heroes of a generation earlier. This cohort grew up in a world in which they could have nearly anything they wanted.

To be sure, pockets of poverty did exist. A considerable proportion of the old minorities, people caught in poverty cycles in Appalachia and elsewhere, did not share in the general prosperity. But probably 80 percent of the people who reached maturity in the early 1970s grew up in a world of extraordinary wealth, by general human standards.

For a second thing, this cohort was the first of the television generations. The oldest among them had begun getting large doses of television advertising by the time they were five and six; the youngest of them could not remember when there was no TV in the house. This was the first group in America to have hammered at them incessantly from early childhood that they must constantly consume. The effect of this barrage is unmeasurable, but could hardly have been minor.

For a third, as this generation was moving into the impressionable teen years, it was confronted by the counter-culture, peopled by youths not much older than they, which was telling them exactly what their television sets had been telling them from infancy: that the *right* thing to do was to indulge themselves, even in what had heretofore been considered dangerous practices. Again, it is difficult to estimate the effect of the hippie movements on the social system, but the figures produced by Yankelovich indicate clearly that the ideas of the hippies very quickly were adopted by the mainstream. By 1971, 36 percent of college students were interested in the idea of communes;[2] in the years around 1970 those who favored laws against drugs dropped from 55 to 38 percent;[3] those opposed to casual sex dropped from 34 to 22 percent.[4]

For a fourth, there was what has been called "the human potential movement," which began in the 1950s and became a real force in social thought in the 1960s. Psychologists like Abraham Maslow, Carl Rogers, Erich Fromm, and many others less well-known were concluding that the potential for happiness, or self-fulfillment, which could be defined in various ways, was much larger than earlier thinkers had believed. Freud had offered what was essentially a pessimistic view of life, which he summarized in *Civilization and Its Discontents,* the main point of which was that the requirements of civilization often operated in opposition to individual human desires and thus left human beings caught in inescapable conflict. The new human potential movement said that many of the institutions of civilization were in fact dead excrescences, residue from the past, which could be shucked off. Steven Mintz and Susan Kellogg say, "Even in the early 1960s, marriage and family ties were regarded by the 'human potential movement' as potential threats to individual fulfillment as a man or a woman. The highest forms of human needs, contended proponents of the new psychologists, were autonomy, independence, growth, and creativity, all of which could be thwarted by 'existing relationships and interactions.' "[5]

The human potential movement was saying, then, that when the needs of the self conflicted with the rules and institutions of the social system, the rules and institutions would have to give way. Fritz Perls, who may have invented the phrase "Do your own thing," even went so far as to advocate "caprice." Says Barbara Ehrenreich, "At its peak, as Joel Kovel has written, the new psychology was both an industry and a kind of secular religion, enlisting hundreds of thousands of middle-class Americans in the project of self-improvement through psychological growth."[6]

Closely tied to the human potential movement—indeed, to some extent caused by it—was the spate of self-help and self-improvement books which reached storm proportions in the 1970s. The *New York Times* best-seller lists for the 1960s rarely carried more than one such book at any given moment, and those tended to be books of presumably practical advice—*How I Made $2,000,000 in the Stock Market* in 1960,[7] *How to Avoid Probate* in 1966.[8]

But the self-help books which began to flood through the culture after 1970 were intended to show people how to live better, richer lives. In 1971, *The Sensuous Man* and *Any Woman Can!* made the list.[9] In the summer of 1973 *Dr. Atkins' Diet Revolution, The Joy of Sex,* and *I'm O.K.—You're O.K.* were on the list together.[10] In 1975 *Sylvia Porter's Money Book, Power! Winning Through Intimidation, The Relaxation Response, The Save Your Life Diet,* and *TM* were on simultaneously.[11] In May of 1978 the following were on together: *The Complete Book of Running, Pulling Your Own String, Adrien Arpel's Three-Week Crash Makeover, Shapeover Beauty Program, The Only Investment Guide You'll Ever Need, Designing Your Face,* and *Looking Out for Number One,* which had at that point been on the list for forty-six weeks.[12] The American appetite for books of this kind seemed to be insatiable, and the effect, inevitably, was to reinforce the belief that a person not only could but should aggressively seek the enhancement of the self.

Finally, the first years of the 1970s saw a sharp break in American morale. This once again is a condition hard to measure, but it is clear that it happened. There had been, of course, opponents to the Vietnam war for sometime, but at the end of the 1960s, especially after the "Tet offensive" of January 1968, it became obvious to the majority of

Americans that Presidents, generals—"the government"—had been lying about the conduct of the war. The general public was also coming to believe that it was a bad war—an unjustifiable one being fought in the wrong way. It was furthermore becoming clear that we would not win the war. Americans had always taken it as an article of faith that the country had never fought an unjust war, and had never lost one. Historians might disagree on both counts, but these were cherished beliefs. And now the country was not only fighting an unjust war, so it appeared to many people, but would lose it.

Then in 1973 came Watergate. Americans had also generally believed that while Presidents might well be misguided, or wrong-headed, they were basically honest and trustworthy. They had refused to blame Harding, Grant, and others for the scandals of their administrations, preferring to think that the Presidents involved may have been lax, or too trusting, but not criminal.

But now the country was faced with a President who had engaged in illegal activities and lied about it. The people were dismayed, and it is significant that there was no popular upwelling of support for Nixon even from his most ardent fans.

Finally, in the late 1960s and the early 1970s, that great prosperity which younger Americans had simply taken for granted began to wither as Vietnam war inflation put an end to the steady increases in real income they had enjoyed.[13] As interest rates, inflation, prices, began to soar, Americans had come again to face a disillusionment about their country. America was, after so many years of such imposing prosperity, beginning to slip.

Taken together, the 1970s produced in the American people a loss of faith in their nation which was, I think, more significant than has been realized. Americans, like most people elsewhere, do identify with their nation. They had always understood that the United States was the greatest nation on earth—the most democratic, the fairest, the most generous, the most prosperous, and since 1945, the dominant power in the world. Needless to say, there were many who would disagree with portions of this assessment—or indeed all of it—but there was enough truth in it to make it believable. Now it was getting harder to accept, and millions of Americans were hurt. This loss of faith, I think, contributed to a certain "the hell with it" attitude, which expressed itself in a mood of cynical selfishness. In any case, whatever the causes, the net result was a dramatic change in the attitudes of Americans, and especially younger Americans. This is, once again, not simply a hunch. In 1957 a team headed by social scientist Joseph Veroff, at the instigation of the United States Congress, made a long and thorough study of American attitudes and feelings to produce what amounted to a psychological profile of the American people. Then, in 1976, his team repeated the study, somewhat expanded. It is only necessary to quote

from the report to see what happened over the course of those some
twenty years:

> A central theme of [the report] concerns what we judge to be a *reduced
> integration of American adults into the social structure.* (Emphasis in the origi-
> nal.) [14]

> [The psychological revolution between 1957 and 1976] expressed itself
> specifically and dramatically in a heightened salience of self-concern. [15]

> These results support Lasch's view that morality has diminished as the
> basis of evaluation and self-evaluation in our society. [16]

> We can say that in 1976 men and women are more inward-looking or
> intra-ceptive. [17]

> The population is more positively oriented toward the self. [18]

> There is a significant tendency for the 1976 population to visit less and to
> belong to fewer organizations. [19]

> There has been a shift from a *socially* integrated paradigm for structuring
> well-being, to a more *personal* or individual paradigm. [20]

This theme—the shift in focus from the social group to the self—
echoes endlessly through the Veroff report, showing itself in a lessening interest in marriage, a greater willingness to accept divorce as a
reasonable solution to problems, less concern for children, and a general sense that social strictures need not be followed if they interfere
with what anyone wants to do. The Veroff report makes melancholy
reading; but it will hardly come as much of a surprise to anybody who
has come this far in this book.

Nor should it come as much of a surprise to anybody to learn that
1973 was a watershed year in American productivity. From 1948 to
1973, "multifactor productivity" rose at the rate of about 2 percent a
year, which meant a doubling of real wealth, roughly, every thirty-five
years. In 1973 productivity stopped rising for almost six years. It picked
up a little in 1979, but was still at only a quarter of the rate of the post-
war years, and it remains at a low level.

Economists have offered a number of explanations for the drop in
productivity, but none of them is wholly satisfactory. According to Edward F. Denison, a senior fellow at the Brookings Institution, "I've looked
for years, and I'd have to say that a good part of the productivity slow-
down can't be explained by anything you can measure. There's something going on now that was not going on prior to 1973." Given what
we now know about the massive shift in American priorities occurring
in the early 1970s, it is not hard to guess what happened.

That brings us to the 1980s. It was thought while the 1980s were still
with us that the United States was witnessing a return to what were

supposed to be the ideas of the Founding Fathers—the old morality of family, church, and community. Few people who have considered it think that anymore: the corporate takeovers, the junk-bond mentality, the looting of the savings and loan industry, and the rest of it make it difficult to believe that a turning back to an older morality occurred. And the statistics bear this belief out.

In fact, the 1980s did see moderate reductions in the nation's intake of drugs and alcohol. One indication of the trend comes from the High School Survey conducted by Lloyd Johnston at the University of Michigan's Institute for Social Research. It shows, for example, that in 1980 some 34 percent of the nineteen- to twenty-two-year-old group said that most of their friends smoked marijuana at least occasionally; by 1987 the number was down to 13 percent.[21] In general, marijuana use among the young, which had doubled from 1975 to 1979, gradually declined through 1988. This downward tendency in drug use among high school students is evident for nearly all drugs.

Cigarette smoking is also down among students, as it is generally in the society at large. And so is the drinking of alcohol. A recent report by the Federal Centers for Disease Control says that per capita consumption has been dropping slowly year by year through most of the 1980s. By 1987, the last year for which figures are available, it had reached 2.58 gallons. (Figures for per capita alcohol consumption vary somewhat from one source to the next.) Most of this decline was due to a substantial drop-off in the use of hard liquor; beer drinking drifted downwards by about 1 percent a year, wine drinking by about 2 percent, on average.[22]

But these welcome declines are in fact not as much cause for optimism as they suggest. For one thing, the incidence of cocaine use in the general population continued to rise through most of the 1980s. For another, the declines are not due to a sense that there is something wrong with abuse of drugs, alcohol, and cigarettes, but that regular use of such substances is dangerous. For example, there was a downturn in cocaine use among high school students in 1987, which Johnston and his colleagues attribute to the well-publicized cocaine deaths of star athletes Len Bias and Don Rogers the previous year. Similarly, the decline in marijuana use is primarily the result of drug education programs and magazine articles reassessing the idea that marijuana smoking is harmless.[23]

Most significantly, the declines, which in some cases are significant, have not brought the consumption of drugs and alcohol back down to anything even approaching the levels of the post-war period. Broadly speaking, we have moved from a peak use of these substances in the last years of the 1970s back to about where we were in the mid-seventies, which was after the sudden upward spurt which came in the early 1970s. What we have seen in the 1980s, then, in respect to drugs and alcohol, is a somewhat slanted plateau; use of these substances is down, but use remains extremely high compared with earlier periods. We are, for example, still drinking more alcohol than we were at any time between the World War I and the early 1970s.[24]

Drugs, furthermore, have established themselves as an American norm: today, by age 27–28, 40 percent of Americans will have used cocaine at least once in their lives.[25] Fifty-seven percent of high school seniors have used drugs at least once, and by the time this young generation reaches its mid-twenties, the figure will be 80 percent.[26] Drinking in colleges has not declined significantly, and in high schools, while daily drinking is generally disapproved of, heavy weekend drinking is taken by at least half of the students as an acceptable norm—for high school seniors, getting bombed on Saturday night is a standard pattern.[27] And in the twenty-three- to twenty-six-year-old group, 51 percent have friends who "do coke."[28] Johnston and his colleagues conclude, "Despite the improvements in recent years, it is still true that this nation's high school students and other young adults show a level of involvement with illicit drugs which is greater than can be found in any other industrialized nation in the world. Even by historical standards, these rates remain extremely high. Heavy drinking is also widespread and of public health concern; and certainly the continuing initiation of large proportions of young people to cigarette smoking is a matter of great public health concern."[29] Today, drugs figure to one degree or another in the lives of perhaps half the American population, and alcohol is a norm in the lives of the majority.

There has been, as well, no return to the sexual morality that obtained in the post-war years. Monogamy is gone as both an ideal and a practice: in one recent year surveyed, about a third of eighteen- to twenty-nine-year-old women had more than one sex partner; among males it was about half. Eleven percent of the women surveyed had three or more sex partners during that year, and among unmarried males some two-thirds had more than one sex partner in the year.[30]

Given this, it is hardly surprising that 40 percent of American children today are conceived before the woman's first marriage. Among blacks, out-of-wedlock births have become the norm: in 1970–74, 67 percent of black children were conceived before the mother's first marriage; by 1985–88 it was four out of five.[31]

One inevitable consequence of the sexual promiscuity which has become a normal behavior pattern in the United States has been the disappearance of the traditional family. Today only 7 percent of the nation's households consist of a working father, a stay-at-home mother, and dependent children.[32] Divorce rates have doubled since 1965; today some 45 percent of all marriages are second marriages for one or both of the partners.[33] Mintz and Kellogg say, "Almost every aspect of family life seems to have changed before our eyes. Sexual codes were revised radically. Today only about one American woman in five waits until marriage to become sexually active, compared with nearly half in 1960. Meanwhile, the proportion of births occurring among unmarried women doubled."[34] So much for a return to family values in the Reagan years.

Americans may be less interested in family life than they were in an earlier day—but they have not lost their zest for keeping themselves amused. They are spending $50.2 billion a year on sports[35]: attendance at major league baseball games went from 30 million in 1975 to 53 million in 1987. Professional football attendance went up 50 percent in the same period, and professional basketball attendance almost doubled.[36] The bill for motion picture admissions went up by over 50 percent from 1980 to 1987;[37] the amount Americans spent for radio and television sets, in constant dollars, was up one and a half times;[38] the amount we spent for cars and car parts rose by some 70 percent between 1980 and 1986.[39] And Americans were spending by 1989 some two billion dollars a year making dial-a-porn phone calls, a business that did not even exist in the post-war period.[40] These rises in expenditures for entertainment have to be looked at against the fact that between 1980 and 1987 per capita income, in constant dollars, rose by 13 percent;[41] in the same years the money we spent for recreation jumped by 42 percent.[42] The priorities of the American people were clear.

Given all of this, it is not surprising that during the 1980s the American people began electing governments that promised to leave them alone as much as possible to do their own thing. Governments reduced taxes, dismantled regulatory agencies and removed other limits on whole industries, fought off efforts to control exploitation of the environment, and even abandoned fiscal common sense so that the *government itself* would be free to do whatever it wanted. To be sure, the government acted inconsistently with its own philosophy when it attempted to require school children to say prayers and tried to interfere with what was now seen as a right to abortion. But its efforts in these directions were feeble and sporadic: what the United States governments of the 1980s offered was the greatest possible freedom for everybody.

Whatever recent governments thought they were doing, their primary philosophy as it worked out in action was built on a reluctance to interfere with the freedom of anybody to do anything. Again and again current administrations have refused to stop corporations from exploiting national resources, like national parks, for private gain; to prevent people from owning whatever firearms they wanted; to place limits on the accrual of unimaginable private fortunes; to require the entertainment industry to curb excessive violence and deceptive advertising in shows aimed at children. Conversely, they have also chosen not to interfere with the right of people to be poor, homeless, sick, and helpless.

Thus, because these governments have insisted on governing as little as possible, they must be measured by what they failed to do, rather than by what they did. And it quickly becomes apparent that, despite their insistence that they were looking back to an older ethic, they were in fact resolutely anti-communitarian and anti-family. They did not do what communities usually do—care for the young, the needy, the helpless; nor did they invest in community projects, like mass transportation, improvement of the quality of air and water, the building of new schools, except to the extent that political realities forced them to. So far as the family was concerned, successive administrations looked with indifference on soaring rates for divorce, unwed motherhood, single parent families. In sum, in the 1980s the United States government adopted as its basic philosophy the ethic of the self: the community and the family were to be sacrificed in order that individuals could grab the best possible deal for themselves.

But this was the sort of government the American people voted for. They knew what they were getting, because the candidates they chose by large margins told them again and again that they would not tax them in order to supply more community services, that they would not provide resources to help families stay together. The American people voted for these candidates because they were offering them what they wanted: as much self-gratification as they could get, and the Devil take the hindmost. And in doing so they made selfishness the *official* policy of the United States.

CHAPTER 18

The Media Over All

The ethic of self, which is today central to the American spirit, has affected virtually every area of life—love, work, sex, child-rearing, politics, religion. One of its most obvious manifestations is the massive involvement Americans have with the entertainment business, which has become for many of us the most significant element in our lives. Nearly all Americans today are engaged with what is called the media for substantial portions of their day. We switch on the radio or the television upon arising; we listen to radio or tapes as we go to work in cars, subways, trains; many who stay at home have their radios or television sets on continuously throughout the day, and the majority of us spend the main part of our evening hours sitting before the television set. Some people, in fact, sleep plugged into a tape recorder: there are people in the United States who are tied to one medium of entertainment or another twenty-four hours a day for weeks, or even months, at a stretch.

The heart of the modern entertainment business is television, in whatever form. Measuring precisely the effect of spending twenty to forty hours a week watching television as most most Americans do, is not easy, but a number of social scientists have been studying the matter; and it is apparent that TV has wrought substantial changes in American lives.

To begin with, TV has come wholly to dominate our leisure time. George Comstock, one of the most respected commentators on the medium says, "There is no more clearly documented way in which television has altered American life than in the expenditure of time. It has not only changed the way the hours of the day are spent, but the choices available for the disposal of those hours, and in so doing has brought the age of mass media to maturity."[1] It is certainly the case that the overwhelming majority of Americans spend more time in front of the television set than they do at anything else except work or sleep. In 1980, the last year for which I have been able to find figures, which in any case cannot have changed much, Americans spent 31 percent of their free time watching television, 7 percent socializing and in conversation, 8 percent reading newspapers, 6 percent engaged with other media.[2]

The significance of this is that it leaves people a lot less time for other occupations. Reading, hobbies, casual lawn sports, card playing, gardening, and dozens of other spare-time pursuits that people used to enjoy are sacrificed to the time-devouring television set. The nature of human life has thus changed for most people, because the experience of watching television is different from most other human activities. As I have said earlier, the process of being "entertained" is a passive activity, lacking the sort of interaction essential to other types of leisure pursuits, like game-playing, story-telling, or just plain conversation. In this sense television watching is, at first glance, no different from radio listening or going to the movies. But, in terms of the amount of time most Americans spend in front of the television set, the quantitative difference becomes qualitative. Going to the movies two or three times a week, as many people did during the Depression, did not soak up the same amount of time; people still had leisure hours in which to garden, hang out on the corner to talk, play badminton on the lawn. Simply by drastically reducing the amount of time Americans spend at more active, engaging, spare-time pursuits, television has changed the nature of life in the United States.

A second important effect of television is that it immerses Americans in fantasy for large portions of their lives. Television shows have a surface resemblance to the real world, for these are after all real bodies we are seeing and real voices we are hearing. But the things that people say and do on television are not the things that people ordinarily say and do; and in most cases they are not the things that people ever *could* say and do. Archie Bunker and J. R. Ewing are just as much mythological creatures as the elves and flying witches and frog princes of the fairy tales. It does not have to be "Star Trek" or "Bat Man": the ordinary soap opera or sit-com is pure fantasy.

And this includes much of what purports to be "news." In an earlier time, television news was mainly in the hands of people who had been trained as "print journalists," and had a newspaper editor's sense of what was newsworthy. Today television news is controlled largely by men and women who were trained in television and are inevitably concerned with how many people are watching. The earlier news commentators, such as Walter Cronkite, were well aware of the tendency for producers of news shows to give big play to events for which there were good "visuals"—that is, dramatic footage of an apartment house blaze with firemen pulling frightened children from the flames, of a train wreck with dazed survivors giving accounts of the experience. Cronkite and others at the time fought this tendency, and would insist on giving time to major events even when there was nothing to put on the screen but a "talking head."

But the competition for viewers has driven news programmers increasingly to feature those stories with strong visuals, with the consequence that fires and train wrecks get a minute or even ninety seconds, while a famine involving millions, for which the only visual available is a map or a still photograph, will get twenty seconds.[3] The final result is that most of what we see on our news programs is deliberately designed to entertain us, not inform us.

Thus, most Americans, who depend upon television for their news, get a badly distorted sense of what is happening in the world around them. To a startling extent television news is as fantastic as "Twin Peaks" or "L.A. Law." Television does have its advantages as a medium of communication. But it is a simple matter of fact that a voter concerned about making intelligent choices, say, can learn more in fifteen minutes with a good newspaper than from an hour of national news. To be sure, there are plenty of papers around that are as bad as television in their choice of what they give their readers; but there are relatively good ones available to most people, too.

Television, thus, is mainly fantasy. The question, then, is, how does this immersion in so much fantasy affect the human psyche? We are far from having a complete answer, but there is already no doubt in the minds of many students of the subject that television has serious effects on people, and most of them are deleterious.

For one thing, there are good studies showing that a portion of the television audience is literally addicted to TV. Once again, definitions of addiction are slippery; but according to one report, "The most commonly used scale to measure television addiction includes using television as a sedative, even though it does not bring satisfaction; lacking selectivity in viewing; feeling a loss of control while viewing; feeling angry with oneself for watching so much, not being able to quit watching and feeling miserable when kept from watching it."[4] By this definition, according to various studies, somewhere between 10 percent and a third of television watchers are addicted to it.[5]

Even those who are not addicted find themselves being drawn to it almost against their wills. Most people have their favorite shows, of course, but as a rule they would rather watch anything than nothing. Even when there is "nothing good on," they will watch. Robert T. Bower has pointed out that although there has been a growing, if slight, dissatisfaction with television programming, watching hours continue to rise.[6]

Why do we watch? According to psychologists Robert McIlwraith and John Schallow at the University of Manitoba, "One common use [of television] is to alter mood. These people turn on the television when distressed. Another is to fill time when you are bored." The numbers who view selectively, choosing only those shows they actively want to watch, are rare, the researchers say.[7]

The truth is that most people find it difficult to turn the set off once it is on. According to Robert Kubey, author of a large-scale study of television viewing, "It's common for people to say they are selective watchers. They'll say they sat down just to watch 'L.A. Law,' but they're still watching three hours later. A great many people feel powerless to get up and turn it off." In part this is caused by "attention inertia," which is "marked by lowered activity on the part of the brain that processes complex information," according to Kubey and his associate Mihaly Csikszentmilhalyi.[8] Thus, in a very real sense, most people for most of the time they are watching television have drifted off into a kind of twilight sleep in which they dream professionally produced dreams. Their mental and emotional systems are in neutral. And it is certainly this lack of engagement that makes the experience of watching television so unrewarding to most people. Kubey says that the more they watch, the worse they feel emotionally. And he goes on to say what ought to be obvious, but isn't: "Our studies also show that human beings are not well designed to enjoy many hours of passivity. Most people feel best, both physically and psychologically, when they are deeply engaged in activities that challenge their skills. And people feel better after being intensely involved in such activities, not worse—as is the case with TV viewing." Television, the investigators say, is not fulfilling: "Only through active engagement with the worlds we inhabit and the people in them can we attain for ourselves the rewards and meaning that makes for psychological well-being."[9]

The group that has been most seriously affected by television is children, for the obvious reason that a five- or six-year-old has more difficulty differentiating fantasy from reality than adults do. According to one 1970 survey, 71 percent of children five years old and younger watch television.[10] Many children are watching enormous quantities of television, and they are doing so because parents do not make much effort to control what their children watch. In this post-war era survey, 30 percent of parents let their children watch whatever they wanted, and 50 to 60 percent exerted only minimal control over their children's viewing habits.[11] It is a startling truth that a large majority of parents think that on balance children are better off with television than without it.[12] This is probably so because, despite everything, the American people as a whole have a high opinion of TV. One authority, Robert T. Bower, says, "Television was perceived, then, primarily as an entertainment medium, offered to the public for their relaxation and enjoyment and embraced by most of them with pleasure and appreciation." They felt that having to watch commercials was a fair price to pay for "free" entertainment. Better educated Americans are frequently fairly harsh in their judgments of TV but watch almost as much as anyone else, and more or less the same shows.[13]

Parents, thus, see nothing wrong with allowing their children to immerse themselves in TV. One recent study of children and television reported, "Children consistently drew themselves smiling as they watched TV . . ." They sit or lie very close to it, often with a blanket, a pillow, a doll, or a pet. "The sense of cozy intimacy with TV was usually reflected in the physical appearance of the viewing area."[14] It is clear from this and other studies that children enter into a relationship with the television set—if permitted to do so—that is quite different from the way they use other types of toys. It has a special meaning for many, if not the majority of them—a meaning more like the relationship with a favorite doll than a machine. They find it, in the simplest terms, comforting, in many cases more comforting than anything else in their lives. They bring it trust and affection.

Unfortunately many of them, especially the younger ones, cannot grasp the fact that most of what they see on television is fantasy—not only unreal, but impossible. They cannot judge it against life, with which they have had only a modicum of experience, and what they see on television looks as real to them as anything else. One consequence of this is that younger children in particular do not understand that television commercials are bought and paid for by people intent on selling them goods. George Comstock says flatly, "There is no doubt that a sizable number of children below some age do not comprehend the nature of a commercial." Comstock suggests that the age, in many cases, may be as old as eight,[15] and he points out that the average child views about 20,000 commercials a year.[16]

Thus, when the nice man or lady comes on the set and urges the child to obtain a toy, he believes that he is actually being offered the item, which the parent only need pick up at the store. When he asks the parent to do so, and is turned down, he often experiences "disappointment, conflict, and anger."[17] The parent suddenly is the bad guy for denying the child something the nice lady has offered him.

Making matters worse is the fact that today many commercials for children are deliberately deceptive. They suggest that toy planes can fly, toy soldiers can walk when in fact they cannot. The people who create these ads know that they are likely to mislead the child, but to date neither the television stations nor any government agencies seem particularly concerned. (In 1990 Congress finally pushed through a bill aimed at placing limitations on television advertising for children, but the limitations are modest.)[18]

What is probably most important about the enormous amount of advertising children are exposed to on television is not that they are gulled into wanting a lot of things that are not necessarily good for them but that they are being told by admired authority figures, from the age of two or three on, when their psychologies are still very malleable, that they *should* acquire things. Once again this is not a matter that has been studied in depth, so far as I can determine, but the effect of it on an unformed mind must be massive. American children are not being urged by their seductive television sets to go outside and play with the other kids, or to save their pennies for something worthwhile, or to shut off the set and do their homework. They are instead being endlessly urged to acquire; and it is therefore not surprising that we now have in the United States a population in its forties and younger which seems to value the act of acquisition for itself. It does not really seem to matter what is being acquired—a new type of soap, a sabre saw, a key chain, a car: the point is simply to go through the acquisition process.

Whatever else can be said about children's television viewing, there is substantial evidence that it does contribute to violent behavior in the young. The point has been debated for decades, going back to the earliest days of television. The response of advertising and television executives has been: (a) the evidence showing a relationship between children's viewing habits and their behavior is questionable; and (b) that it is likely that television violence has a "cathartic effect" by which violence-prone kids let off steam.

Yet the social scientists who have tried to study the matter with carefully contrived experiments have shown again and again that, despite what the publicists for the television stations claim, violence on TV does produce violence in children. George Comstock says, "By . . . the beginning of the 1970s, about 50 experiments [demonstrated] that children and college students were likely to display increased aggressiveness immediately after viewing a violent portrayal."[19]

And this was true not just of American kids but of kids everywhere. For example, experimenters in Israel found that "among city children a significant positive correlation was found between television violence viewing and amount of aggressive behavior. . . Further, longitudinal effects seemed to be more from violence viewing to aggression, than from aggression to violence viewing."[20] In other words, it was not that violence-prone kids tended to watch violent television shows, but that violence on TV made many children want to do something violent themselves.

Why are we so concerned about television in this respect? There has been violence in "the media" dating back to the *Iliad,* the myths of King Arthur and Robin Hood, Shakespeare, the movies, novels, comic books, adventure magazines. (In particular there exists today a genre of violent comics which are deliberately designed to appeal to a sadistic impulse.) The answer is threefold: for one thing, children spend far more time watching television than they do reading Shakespeare, or even comic books. For another, the amount of violence packed into a half hour of television is far greater than that in a half-dozen plays of Shakespeare, any ordinary novel, or adventure story. Finally, television has a terrible reality which cannot be matched by fiction, radio, oral tradition. These are real people firing real guns, stabbing with real knives to produce what appears to be real blood. There is no act of the imagination involved. It all seems like life.

But of course it is not: it is always fantasy. And here, primarily, is where the problem lies. The violence on television is presented as if it were part of the fun. The maimings and deaths are not meant to provoke in us the feelings of terror and disgust they would occasion in real life. They are given to us as larks, acts without consequences, meant solely to keep our attention focused on the set. There is no shrieking widow, no bereft child traumatized by the shooting of a father, no young man doomed to spend the rest of his life in a wheel chair. Television violence is not real; it is fantastic. But it seems real, and it can leave the viewer, especially an unthinking and half-formed child, with the idea that violence is not serious; that it does not matter very much if you hurt somebody. And that, say people who have studied it, is precisely what it does.

If people wish to spend substantial parts of their lives in the semitrance that most television produces they have, in a democracy, a perfect right to do so. What is troubling about this enormous immersion in television, and the media in general, is the extent to which it has isolated people from one another. At bottom, television is a machine which helps people to wall themselves off from one another. So long as we are engaged with the magic box, we are not engaged with others. What does it mean for a parent to sit close to a child when both are more aware of the car chase on the screen than they are of each other? They are to each other simply warm bodies—isolated selves who have shut themselves up in a fantasy world the box has made for them.

CHAPTER 19

The Consequences
of Selfishness

If the passion for dedicating one's life to one's self which is now endemic in America had no serious consequences, there might be little objection to it. After all, why should people not enjoy themselves as much as possible in their brief lives on earth?

Unfortunately, the deep concern for the needs of the self which grips most Americans today *does* have serious consequences that are already manifest in the society, and are certain, in my view, to damage the social system and to cause pain to the people who make it up. There are, it seems to me, three areas of American life in which the heightened selfishness of the present day not merely has the potential for causing substantial harm, but already has.

The first of these is the extent to which we have abandoned our children. Between a soaring divorce rate and an equally soaring rate of children born to unwed mothers, it is now the case that the majority of our children will spend at least a portion of their childhoods in single parent homes—in effect being raised without fathers. A large minority will spend their entire childhoods essentially without fathers, and a considerable number will not even know who their fathers are.

This is an extremely unusual circumstance—perhaps unique in human experience. *In no known human society, past or present, have children been generally raised outside of an intact nuclear family.* The nuclear family is one of the most basic of all human institutions, a system of doing things so fundamental that until this century it occurred to very few people that life could exist without it.

There has been, of course, a good deal of variation from one culture to the next in the way the family works. In some the role of the tribe, or various relatives, in socializing children is greater than in others. In some groups mothers and their offspring form sub-clusters, with fathers less in evidence than they are elsewhere. Again, as for example among the Haida Indians of the Northwest in an earlier day, boys are removed from the family for several years to be raised by their uncles.[1] For medieval Europe the sons of busy monarchs often were put in the charge of respected senior knights;[2] and in England even today it is the custom for people of the upper classes to send boys, and sometimes girls, away to boarding schools at the age of eight or so.

But everywhere the nuclear family has the fundamental responsibility for the care and training of the offspring. *Nowhere* is the father wholly absolved of the duty of supporting his children, nor in most cases of supplying some measure of guidance, emotional support, discipline, and nurturance.

In recent years it has come to be widely believed in the United States that a single parent can raise a child as successfully as two parents. For one thing, the single parents among us—I was one myself—dislike thinking that we are doing less well by our children than we could have had we stayed married. For another, it had been important to many in the women's movement to believe that women can operate independent of men, even in respect to the raising of children. As a consequence we have been lapping up a slew of popular books on single-parenting which take an extremely optimistic view of the subject: *How to Single Parent, Single Father's Handbook, Sex and the Single Parent, Succeeding as a Single Parent, Successful Single Parenting, Enjoying Single Parenthood.*

Unfortunately these books have obscured a host of scholarly research on single parenting which takes a much less sanguine view. The serious literature on single parenting, some of it going back to the 1960s, is enormous. Henry B. Biller, a leading authority on homes without fathers, calls the outpouring of information on father-absence "a dramatic explosion in data."[3] Biller himself, in a recent book, has reviewed over *a thousand* such studies.[4] The evidence is overwhelming: children growing up in single-parent homes—and that really means without fathers—consistently do less well on any measure than children from intact families. This is today quite simply beyond dispute. Socially, intellectually, at school, at play, children of divorce do not function as well as do kids with their fathers at home. "Well-fathered infants are more curious in exploring their environments,"[5] Biller says. They are more secure, more trusting. Their motor development may be more advanced.[6] Carefully controlled studies have linked father absence with tendencies toward delinquency, confused gender identity.[7] Boys and girls from father-absent homes are likely to test lower in "perceptual-motor and manipulative-spaecial tasks."[8] Many tests have shown that kids from father-absent homes scored lower than those from intact families on "social knowledge, perception of details and verbal skills."[9]

Even as college students these young people score lower on verbal, language, and total aptitude tests.[10]

Fathers are particularly important in helping both boys and girls in establishing good sexual identities. E. Mavis Hetherington, in a series of important studies, has shown that daughters of divorce are likely to be sexually aggressive as teenagers, are more anxious, have a greater sense that they lack control over their own lives. She ways, "Women from intact families tended to make the most realistic and successful marital choices."[11] All of this suggests that it is not enough for a girl to have a good "role model"; for the developing child there is clearly a need to interact with a male figure.

Furthermore, a number of studies show that mothers without husbands at home are more likely to abuse their children than married mothers. Biller says, "The high level of maltreatment of children by single mothers can also be viewed, at least in part, as being related to a social system that tends to put too much pressure on the mother's accountability and not enough on the father's participation in child rearing."[12] A husband who provided his wife with emotional support has a real, if indirect, effect on the child. "In fact, even before the birth of the child, the presence of an emotionally supportive husband can contribute to the expectant mother's sense of well-being and is likely to be associated with a relatively problem-free pregnancy. [One study] revealed that mothers were more interested in caring for their newborn infants, including nursing them, when they received a high level of emotional support from their husbands . . . The quality of the husband-wife relationship is more predictive of the mother's success in dealing with her child than is the degree to which the father actually participates in child care."[13]

None of this suggests that every divorce is inevitably disastrous for children. Biller says, "Although we advocate the advantages to the child of living with two parents, we do not automatically depreciate alternate family forms. For example, some children develop very well though one parent is frequently absent from home as a result of occupational demands; others thrive in so-called joint or shared custody situations after divorce or in the presence of a highly effective single parent."[14]

Indeed there are at least a few situations in which children are better off without their fathers—when they are abusive, ineffective, or passive, as Mavis Hetherington in particular has shown.[15] But we are speaking here of relatively extreme cases. In general, the critical factor for the children of divorce is the continuous, supportive presence of a father on something like a daily basis in the lives of the kids. That is to say, the more the arrangement resembles a two-parent home, the better the outcome is likely to be. Thus, given optimal conditions the children of divorce can do well.

But conditions are rarely optimal. Few divorces are entirely amicable, and hostility between parents makes it difficult for fathers to remain within a family circle. Remarriage of either partner brings an unpredictable new element into the situation, especially if there are step- or half-siblings. The careers of fathers especially may draw them away from their children. And then there are always those fathers who are uninterested in their children, or, to put it in the context of this book, more concerned with their own self-fulfillment than with maintaining frequent contacts with their kids. A step-parent may be a wonderful person but, as anyone who has been a step-parent knows, the role is filled with tensions and difficulties. Says one recent report, "Studies of divorced fathers show alarming numbers retreating from emotional and financial commitments after a divorce. By some estimates, nearly half of all children in the custody of their mothers have little or no contact with their fathers." Andrew Cherlin, a sociologist from Johns Hopkins who has studied the matter says, "There's a relatively small group of well-educated and sensitive fathers who are prevented by custody arrangements from acting as responsible parents. But their numbers are dwarfed by those who don't give a darn."[16]

Henry Biller concludes, "Both men and women have something special to offer children. For various important cognitive, emotional, and social learning experiences, the presence of two parents is advantageous for the child. . . . The lack of adequate fathering does not necessarily lead to psychopathology, but it can certainly impede the actualization of the child's talents . . . Children develop best when given the opportunity to form a basic relationship with a positively involved father and a positively involved mother. Fathers are as important as mothers in the overall development of children."[17]

The bottom line, then, is that children of divorce can develop well, but the odds are heavily against it, simply because there are too many things that can get in the way of good parenting—the lack of money, demanding careers or jobs, complicated sets of feelings, remarriages and all the rest that anyone who has been divorced knows far too well. What this means, finally, is that, contrary to what is almost universally believed, it is *not* better to divorce than to raise children in a home with tension and bickering. This statement will seem to many people contrary to common sense; but there is plenty of evidence that children are better able to deal with a pair of bickering parents than even the partial loss of one of them. Yes, in the extreme cases where fathers are abusive, bullying, and hostile, it is probably better to remove the children from them. But these are only the extreme cases. It is not enough that a father has a quick temper or is demanding of his children to make divorce preferable: all parents become irritated by their children and shout at them from time to time, even fairly frequently. But for most parents the old and today discredited idea of staying together for the sake of the children is worth thinking about.

It is also in this context worth thinking about the abandonment of the strictures against pre-marital intercourse. Very rarely in human society have people been allowed to breed outside of marriage. Ira Reiss says, "The key societal function of marriage is to legitimize parenthood, not to legitimize sexuality. Sexual expression is common outside of marriage in the preliterate world. However, because marriage legitimatized parenthood, premarital relations that eventuate in conception without the probability of marriage are almost universally condemned. Thus, most societies are more tolerant of premarital intercourse between engaged couples, for they are likely to marry if pregnancy results."[18]

Other reasons have been advanced for this universality of hedges around sex outside of marriage. Some have said that it assured that fathers knew their children were their biological descendants. Others have said it was to preserve male property rights in wives and daughters. However, the most obvious reason for limiting premarital intercourse is to ensure that children will be raised by two adult parents. A major result of the promiscuity of the young has been to produce a huge crop of children being raised by immature women, often without the help of fathers.

Exacerbating the problem children face in our society is the fact that both divorced mothers and mothers who never married in the first place will almost certainly have to work part-time, and in most cases full-time. Today, 57 percent of women with children under six work; in 1950 it was 12 percent, according to one study.[19] Another study puts the figure higher, saying that half of the mothers of children under three work, 60 percent of those with children aged three to five.[20] Thus, the child of divorce has a part-time mother as well as an occasional father. One study of about 5,000 middle-class eighth graders in Southern California showed that 28.6 percent of so-called latchkey children were unsupervised most of the after-school hours—in itself a dismaying figure—and that these latchkey children were more than twice as likely as more supervised children to drink, smoke cigarettes, or use marijuana.[21]

The consequence is that today the majority of children will spend significant portions of their childhoods away from their parents, from early ages. Even as far back as 1982, according to the U.S. Census figures, only 30 percent of preschool children were cared for in their own homes. Forty-one percent of them were in another home; 15 percent had group care; 14 percent went to work with their mothers; and 5 percent were unaccounted for—thousands of children simply left without adult supervision.[22]

This shifting of so many children out of their homes to be raised by surrogate parents again is a social change away from the human norm of a considerable magnitude. The experiment has been tried before, as for example in the kibbutzim of Israel, and in various socialist nations, partly out of the belief that children could be better socialized by professionals than by their own parents. But in neither of these cases did the children being raised in groups constitute a majority of the population, or anything like it. In the United States, if present trends continue, it is probable that by the end of the twentieth century almost all American children will spend the bulk of their childhoods in day-care centers of one kind or another, seeing their parents only for two or three hours in the evening, and for the portions of the weekends when neither parents nor the children are occupied with tennis, dancing schools, and Little League practice.

Already in this society millions of children are spending fewer than twenty waking hours a week with their mothers—less time than most of them spend watching television—and those hours are spent with a harassed woman who is trying to get one or two small children off to day-care and herself off to work; or a tired woman at the end of the day wanting simply to feed the children and get them to bed as expeditiously as possible, so she can sit down for a few moments in front of the television screen before it is time to fix tomorrow's lunches.

American children today are spending twice as much time in institutions as in their families. The question is how well do these institutions work? The jury on day-care is still out. The first studies were hopeful: they seemed to show that day-care children were doing well, and might even be more advanced in certain ways than children who stayed home with their mothers, particularly in being more assertive with their peers, and even in some aspects of their cognitive development. A recent review of the literature reported, "In general, the reviewers have concluded that good quality day care does not appear to have any adverse impact on most aspects of young children's development."

Additionally, it has been well-demonstrated that children from disadvantaged homes, where there may be some combination of poverty, father-absence, alcoholism or drug-taking, will almost certainly benefit from day-care.[23]

But many of the social scientists who have been studying day-care are left feeling somewhat uneasy about it. They are particularly worried about the effect removing children from the home will have on the mother-child bond, and the long-term consequences which might follow from the disruption of this bond.[24]

We are a long way from understanding how this kind of bond works, and what its importance is in the rearing of children—indeed, in human life in general—but many social scientists are reluctant to simply dismiss a pattern of behavior so ubiquitous to life.

Furthermore, some social scientists have pointed out that the bulk of the day-care centers which have been studied are the high-quality ones associated with universities, which may be atypical, although one study

of a more typical day-care center in a North Carolina community concluded that these children were doing just as well as home-raised ones.[25]

But the doubts remain. A number of studies have suggested that
day-care children form less secure attachment to their mothers, and
presumably to their fathers as well, although the evidence is to some
extent ambiguous.[26] There is fairly widespread agreement that day-
care kids are likely to be more aggressive than home-raised ones. They
"interact more with peers in both positive and negative ways," hardly a
surprising finding. The report adds, "Children with day-care experience are less likely than home-reared children to fully conform to adult
standards of behavior and comply with adult requests."[27] Marian Blum,
a researcher who has raised questions about day-care, says, "Until more
longitudinal [i.e. long-term] studies are completed, there is no way of
knowing what long-term, so-called sleeper effects may be due to attendance at a day-care center."[28] And a team which reviewed the recent
literature concluded, "There are enough negative and contradictory
findings in the literature, however, to raise persistent questions about
the effects of day-care, particularly in relation to social and emotional
development."[29]

Thus, while there is no conclusive evidence that day-care harms children, enough doubts have been raised to suggest that we do not yet
know enough about day-care to accept it as an alternative to traditional
parenting. One important factor, most researchers agree, lies in the
quality of the care.[30] It has been found that small units, in which the
care-givers can talk regularly to the children, help. The problem lies in
the fact that most people who work in day-care centers are underpaid
and overworked. Staff turnover is high—as much as 40 percent annually in day-care centers, 60 percent in family day-care homes[31]—the
consequence of which is that children in day-care are again and again
confronted with the loss of surrogate parents, and with strange new
faces.

Moreover, day-care has become a booming and very profitable business. Inevitably day-care entrepreneurs are as much concerned with
profit as they are with giving the highest quality of care, and compro-
mises are certain to be made. The question of whether day-care—the
raising of the next generation of Americans—should be in the hands
of private business is an extremely serious one. It has generally been
believed that the education of children ought not to be a profit-making
business but one aimed at providing the best education the community
can afford. It is hard to see why day-care for younger children should
be any different.

Between the high divorce rates, the rising numbers of children born
to unwed mothers, and the widespread institutionalizing of young children, we have seen in America an abandonment of parental responsibility which is unmatched in human history. There remain, of course,

millions of caring, concerned parents struggling to raise their children the best way they can. But this group is the minority. It is probable that we are now raising a generation that will be less well socialized, more self-centered, and probably somewhat more impoverished in its cognitive functions than previous generations. The damage, I submit, has already been done, and the results are abundantly evident in the rates for crime, alcoholism, drug use, and disaffiliation we are seeing in our young.

It has been argued that there is no help for it: millions of mothers are single parents willy-nilly, and other millions must put their children out to day-care because they have to work. Neither argument will wash. Single mothers are single because (a) they chose to have babies without first providing fathers for them; or (b) because fathers decided to seek greener pastures at whatever cost to their children, their wives, themselves, and the social system as a whole. In other words, parents today are making *the choice* to have children without accepting any concomitant responsibility for taking care of them. It is particularly the fathers who are to blame. Millions of American males are today refusing to accept any responsibility for the children that they have sired; and even more millions are escaping from those responsibilities as soon as they discover that child-rearing is not always fun. It is more important for perhaps the majority of American men to fulfill themselves than it is to see that their children get off to the best start in life. It is difficult to have much sympathy for their complaints about the child support payments they are asked to make.

Nor will the claim hold up that times are so parlous that it takes two incomes to support a family today. Per capita disposable personal income in constant dollars—that is to say, discounting for the effects of inflation—is more than *twice* as high as it was in 1950, and *three* times as high as it was in 1930.[32] That is to say, the real income of today's parents is double that of their own parents. Young fathers and mothers will complain that housing costs are much higher than they were in 1950, which is of course true. But even when inflated rents and house prices are factored in, *today's parents can live better on one income than could parents of a generation ago.*

But they do not choose to do so. What parents today have forgotten, or did not ever realize, was that their parents accepted the fact that they would start their families in cold water flats, in tiny apartments over garages, in attics over in-laws' homes. They forget that many of their grandparents raised families in cramped tenement apartments heated by coal stoves, often with toilets in the hall shared by everybody on the floor. They forget that their great-grandparents were raising children on farms without electricity, running water, or central heating—or indeed, were doing their "parenting" in hovels on the Hungarian plains or in the desperately impoverished villages of Sicily.

Today's parents, however, were raised to that astonishing prosperity we have looked at. They have come to believe that it is *necessary* for them to have new cars, VCRs, camcorders, frequent dinners out, trips to Aspen, the Bahamas, Europe. But these things are not necessary: having them is a matter of choice. Most fathers today, even those working at relatively low level jobs, can earn enough to support their families better than their parents did. Those parents did not take expensive vacations; they did not eat in restaurants except as birthday treats or anniversary celebrations; they did not own new cars, and many of them did not own cars at all. They had to scrimp and save to pay for the new television sets coming on the market; it never occurred to them to spend money on drugs, and even the purchase of a new suit for Dad was a matter of considerable family debate. These people may have wanted all those things, and envied those who had them. But they had been brought up to believe that children ought to have their fathers living with them and their mothers at home until they were old enough to fend for themselves; and in 70 percent of cases they were.

Nor will it do to argue that women are entitled to careers and must perforce take minimum time away from their jobs in order to rise in the work place. The elitist assumption that most women have, or even could have, "careers" is simple nonsense. Doctors and lawyers together constitute about 1 percent of the American work force. Writers, artists, entertainers, and athletes constitute about another 1.5 percent of the working population, and that figure includes huge numbers of young musicians, free-lance journalists, and unemployed comedians who are barely getting by and will soon find themselves in more mundane careers. Taking it at its furthest stretch, at most 25 percent of American workers are on what could be called a career track.[33] Three-quarters of Americans must work at ordinary jobs as file clerks, stockroom personnel, truck drivers, mailmen, sales people, and the like, and almost half still work in traditional blue-collar jobs as waitresses, carpenters, bus drivers, freight handlers, farm hands, and machinists. It is probable that 80 percent of mothers can easily move in and out of the work force without much loss of pay, seniority, and other perquisites, simply because most available jobs do not offer seniority or perks.

It is not often pointed out in the debate over rights of women to careers that most men do not have careers either. In fact, those men who do take responsibility for their families have usually had to give up a dream in order to take the boring and unchallenging jobs most people work at in industrial society. They will never become trotting horse trainers or novelists or pit mechanics or poets or croupiers or philosophers. Instead, if they are to provide for their families, they will spend their lives hammering dents out of fenders, stocking supermarket bread shelves, checking bills of lading, welding steel beams. Some time ago a friend happened to be with a teenaged son of the 1960s generation when a commuter train passed by, filled with home-bound husbands wearily reading their evening papers. The son announced, "Boy, I'm never going to be like those guys." My friend replied, "Those men are heroes." And he was right.

Raising children is not easy. It is often physically tough, and always emotionally demanding. But raising children is the most important thing any social system does, for if it does not produce a competent, concerned new generation, it will shortly cease to exist. I do not expect that American society will die out immediately. But so long as Americans continue to put the interests of themselves over the needs of their children we are going to create a social system which each year will be less pleasant to live in.

This is not simply theoretical. An extremely significant study by Douglas A. Smith and G. Roger Jarjourq of fifty-seven neighborhoods comprised of 11,419 people shows that crime is not associated with race or poverty; it is associated with single-parent homes.[34] That is to say, the high crime rate in our ghettos is primarily due to the fact that a very high percentage of children there—the majority in many cases—is being raised without fathers.

Another study shows us why. Dr. Richard Koestner, a psychologist at McGill University, and his co-workers followed a group of women and men who had been studied as children in the 1950s. They were astonished to find that a primary—and perhaps the primary—force in producing empathic adults was the presence of a concerned father in their childhoods. "We were amazed to find that how affectionate parents were with their children made no difference in empathy. And we were astounded at how strong the father's influence was after 25 years."[35] It was not what fathers did; it was simply the fact of their being there that produced more empathetic offspring. People who cannot empathize with others tend to see them as if they were trees or stones, and thus find it easy to prey on them.

I believe that the abandonment of the children is the worst of the consequences of the culture of selfishness, but running not far behind, and related to it, is our insistence on having huge quantities of private goods at the expense of public needs.

This is hardly a novel idea. John Kenneth Galbraith made the point specifically in his 1967 book, *The Affluent Society*.[36] Galbraith's conclusion was that we were putting far too high a percentage of our resources into the private sector, and far too little into the public sector. To put it in concrete terms, we were spending too much on television sets and not enough on schools; too much on hunting weapons and not enough on police; too much on automobiles and not enough on mass transportation.

I do not wish to get embroiled in the argument over whether a budget deficit the size of the present one is a danger to the economy, or over the risks of sales of American assets to foreigners, or similar questions. Even without considering the size of the national debt, trade deficits, and the amount of money Americans owe on credit cards and mortgages, it is abundantly clear that the country is facing substantial, serious problems that are going to require huge sums of money to solve. I do not believe that money alone can solve all our troubles: it will take intelligence and imagination as well. But it will take money, a lot of it, too. We clearly need more money going into our schools, more into the day-care centers that now substitute for parents, more into bridges, roads, aqueducts, and the rest of the infra-structure essential to the running of an industrial society. We are faced with a massive bill for repairing the environment which industrial societies, with their profligate uprooting of forests, drowning of rivers in filth, and ripping mountains down to get at the minerals inside, led the way in damaging. We need billions of dollars' worth of drug programs, billions of dollars' worth of prisons and vast new law-enforcement systems. Whatever your priorities—crime, drugs, the environment, education, the homeless, the deficit, the handicapped, the aged, national defense, space exploration, liberal, conservative, middle-of-the-road—they are going to take a lot of money to achieve.

This means paying substantially higher taxes. We as a nation have failed to grasp the idea that a modern industrial society cannot be operated by private business. A huge proportion of what needs to be done to keep the system running cannot be done for profit. Schools, mass transportation, police forces, the judiciary, the armed services, fire protection, parks and playgrounds, and much else have to be operated and paid for by governments, simply because we must provide these services to everybody, regardless of their ability to pay. Imagine what sort of society we would have if all roads were toll roads, all police and fire departments were private concerns answering the calls of only those well-off enough to subscribe to them, all schools were private so that the wealthy would have good schools for their children and people with lower incomes poor ones for theirs. The idea is incommensurate with democracy; it may well be that other functions ought to be taken out of the hands of private concerns: as I have suggested, if day-care becomes solely a profit-making industry, we are certain to have one quality of day-care for the well-heeled, another quality for the ordinary wage-earner.

What we must come to understand is that taxes are not a punishment inflicted on us by an unfair government but buying choices which are available for Americans to make. Do we need more ski vacations—or better police forces? Do we need three or four television sets in each home—or better provision for the homeless? Do we need more power boats, more recreational vehicles—or cleaner air? It is not necessary to belabor the point.

If Americans were being ground into the dust by an oppressive tax collector who was stripping them bare and leaving them to cover their nakedness with a barrel held up by suspenders, as the cartoonists used to have it, we might be right in trying to fend off new taxes. But that is not the case. Since 1980, in constant dollars, purchases of goods and services by the public have increased 26 percent faster than the growth of the economy as a whole.[37] This buying splurge has been fueled by a doubling of consumer debt between 1980 and 1987—from $300 billion to $600 billion.[38] A huge percentage of this money was going for entertainment and what can only be called toys—recreational vehicles, sporting equipment, electronic gadgets—tens of billions of dollars' worth. As a whole, through the 1970s the amount we spent for amusement and recreation jumped by 50 percent; between 1980 and 1987 it climbed by another 48 percent—a doubling of what we spent for fun—in real dollars—in seventeen years.[39]

By comparison, during the 1970–86 period, the construction of new houses remained level, despite the desperate shortage of housing and the considerable increase in population.[40] Over the same period we increased the acreage of our national forests by only 2 percent,[41] in the face of the obvious necessity the world now has for trees to maintain our present ecology. There are actually fewer hospitals in the United States today than there were in 1960,[42] despite the fact that the population had increased by about a third in that period. Nobody has to be told that roads, bridges, sidewalks, and much else in the big city infrastructures have deteriorated badly. These examples of neglect of the basic machinery of the American culture can be multiplied almost indefinitely.

We have, in my view, been making poor spending decisions. What we have been doing, in essence, is electing to buy the cheapest possible education for our children, the smallest possible police force for our cities and towns, the worst kind of highways for our countrysides, the most minimal water supply systems for everybody.

We are not a destitute nation. According to Henry J. Aaron, a senior fellow at the Brookings Institution, "We are an extremely rich country. The question is whether we are raising enough to cover the public expenditures that bipartisan majorities insist the nation requires. And we aren't." Aaron goes on to point out that "taxes as a percentage of national wealth are lower in the United States than in most other industrial countries, and that the amount of national income going into taxes has remained essentially unchanged in this country for two decades,"[43] despite the obvious and acknowledged need for drastic improvement in schools, child care, law enforcement, mass transportation, the environment, and much else. We have the money; but we are spending it on new wallpaper and draperies, instead of repairing the cracks in the foundation and the leaks in the roof.

Look at some hard figures. Recently the government has spent a year struggling mightily to find about $10 billion to finance the war on drugs. In the same period the American people spent $21 billion—more than twice as much—on pizza.[44] In 1986 all government—state, local, and federal—spent $44.3 billion for health and medical programs;[45] in the same year Americans spent $50 billion on spectator sports, and that does not include the enormous sums spent in gambling on sports. In 1987 the United States spent about $170 billion on public elementary and secondary schools;[46] in the same year the citizens of this great country spent almost as much—an estimated $150 billion—on drugs.[47] Once again such examples can be strung out indefinitely. What they tell us is that there is plenty of money out there. And they show us further that, because there is so much money around, we could find substantial sums for taxes without much sacrifice to our well-being. Recently I was told a story of a man living in a well-to-do Connecticut suburb who was one of the leaders in a fight against an increase in the local school budget. The school principal told my informant, "The new budget is going to cost that man about $100 a year. He has three kids in my school and I know for a fact that each one of them has a pair of $175 running shoes."

The money is there for cleaner air, more reliable subways, larger drug rehabilitation programs, and all the rest of it—if we want these things. But we will have to give up something to get them. It is a very simple equation: parents who vote for candidates who promise to hold down taxes are ensuring that their children will have available to them a rich smorgasbord of drugs to choose from. Big-city dwellers who vote for such candidates are making it certain that murderers will be released back onto the streets from overcrowded prisons. Suburbanites who vote down town budgets and shriek in agony when the local school tax rises by 10 percent are guaranteeing that the problems of the inner cities which they fled will presently come out of the ghettos and camp on their doorsteps. And it follows that politicians who attempt to persuade the people that their needs are best met by providing them with poor schools, dirty subways, and inadequate police protection in order that they may buy trivialities are doing them an immense disservice. Paying taxes may be painful, but it is not nearly as painful as being mugged, or seeing a beloved child in the grip of a drug addiction.

This insistence, then, on spending our wealth on self-indulgent trifles, rather than for goods and services basic to the health of the society, is, in my view, the second of the serious problems the cult of selfishness has brought with it.

The third issue that concerns me is less susceptible to measurement, but is nonetheless, I believe, palpable. That is the opinion, which seems to be held by millions of Americans, that they have a constitutional right to break the law. Not long ago a citizen of one big city—me—

admonished a young man who had failed to clean up after his dog, as required. The young man's response was, "It's none of your damn business," or some such words. He felt that whether or not he obeyed the law was a matter which concerned only himself and some vague "them" in authority. He was unable to understand that there was a purpose to the law—that it existed for a good reason.

Most of the uncounted thousands of laws, rules, regulations under which we live are reasonable and sensible. Only a very few, like laws concerning abortion, gun control, the environment, and a few other matters, are subject to much debate. There are good reasons for laws against eating on buses, playing loud music late at night, parking on sidewalks, emptying sludge into rivers, selling liquor to minors, battering children.

Yet substantial numbers of Americans feel that they have a right to speed up as the light turns yellow at a crossing and dash through the red, in the process every year killing a considerable number of pedestrians. Bicyclists in big cities take it for granted that they can weave through traffic in the wrong direction, regardless of the risks to themselves and the pedestrians they frequently knock down. Millions automatically falsify their tax returns—a Harris poll concluded that 53 percent of "yuppies" did their own taxes because they could not find an accountant who would sign them, so blatantly did they cheat.[48] Millions more daily break the laws concerning drugs, littering, air pollution, stealing, smoking, noise-making. Even the common courtesies are ignored, as people push aside others in order to jam themselves first into subway cars, play their radios in public places, put their feet up on railroad coach seats.

Far too frequently Americans believe that the law simply does not apply to them—that they somehow have a right to live unconstrainedly. Needless to say, a culture in which even a minority feels that it has a right to do what it wants cannot possibly be a happy one. But that is what *the majority* in the United States believes.

It is in the nature of human life that people will want things; but only small children believe that they should have everything they want—or so it used to be. A nation in which most people cannot even occasionally put the good of the whole society above their own immediate gratification is bound to grow steadily worse.

I have been talking to this point about what the cult of the self has done to the society as a whole; but there is another aspect to it, and that is what this unceasing concern for the self has done to *itself*. For it appears to me that selfishness does indeed harm the individual in ways we are only beginning to recognize.

One of the significant conclusions of the Veroff study was, "There is a significant tendency for the 1976 population to visit less and to belong to fewer organizations" than had been the case with earlier generations.[49] According to these figures, fewer than a third of Americans visited friends and relatives "more than weekly."[50] Anybody growing up in an earlier America embedded as he was in friends and family, would find this fact fantastic. Even people born earlier in this century will find it startling that most young Americans are interacting so infrequently with people they presumably care about. But among those same people, the number who were married dropped by 16 percent between the mid-1950s and the mid-1970s;[51] the number of children they had at home had also gone down;[52] and the number of single parents had soared.

As we saw much earlier in this book, in the seventeenth century it was actually forbidden for people to live alone. By the eighteenth century it was permissible, but considered an oddity. In the nineteenth century, with the development of the industrial city, increasing numbers of people, especially the young, were living by themselves in boarding houses; but nonetheless they constituted a small fraction of the population and, furthermore, they assumed it was a temporary condition.

Today, living alone is commonplace. And it is particularly the younger generation which is doing it: between 1970 and 1987 the number of Americans living by themselves doubled, and in the twenty-five- to forty-four-year-old group the number quadrupled.[53]

In sum, we are becoming a nation of loners. Increasingly younger people reject marriage, divorce easily, abandon their children, have fewer friends and see less of them. The figures very strongly support the conclusions of the Veroff study that there is simply less contact between people today than there was in a past time.

How do we explain this? In part it may have to do with the intense involvement with the media, which provides a substitute for human interaction. In part it may have to do with increasingly more casual parenting, which gives children less experience with intimacy from early ages and probably creates fear of it. In part it may have to do with the great prosperity of the nation, which allows so many people to have their own apartments, to divorce, to buy all those expensive pieces of equipment that provide substitutes for human contact. But at bottom, it seems to me, the increasing fragmentation of the American people is a consequence of the long-term turning inward to the self as the primary concern of life.

This focus on the self is not without cost. One major finding of the Veroff study was that "in 1976 young people report increased symptoms of psychological anxiety and a greater frequency of worry than did the young in 1957."[54]

It is my view that the two go hand in hand. As I have said, the environment in which human beings are meant to live is one *with* other human beings. If, for whatever reason, we lack close contact with other people, we are certain to feel loneliness, depression, anxiety. Conversely, we will not feel that full sense of well-being which comes from being with friends, family, a social group. Philosophers have often said that the best way to find the self is to lose it: that is to say, people are frequently at their happiest when they are absorbed in something outside themselves—a task, a child, a game, a lover, a hobby. And surely the most selfless of occupations is involvement with others.

This, of course, is precisely what the Kubey study of the effects of large amounts of time spent before the television set is telling us: people who do not interact regularly with other people are likely to lose "the rewards and meaning that make for psychological well-being." The American social system is today fragmented. Husbands and wives, parents and children, friends and relatives see each other only occasionally, and in passing. The old idea of the closely knit family, the tight community, is gone. To be sure, it was not always heaven inside of such human groups, for there was always friction and constriction in them. But there was the pay-off, too, in the sense of security such groups provided, in the warmth of feeling that people who live and work together often share. As Kubey and his colleagues say, the human being was not constructed to live without steady contact with other people, and those who attempt to do so will suffer for it, no matter what.

It is important for us to bear in mind Justice Holmes's famous dictum that your right to swing your arm ends at my nose. Nobody's freedom is unlimited; everybody's rights are curtailed by the needs of others. Any involvement with other people as friend, competitor, spouse, parent, or other relationship limits the freedom of both parties. But too many of us fail to understand this, and in the pursuit of an unattainable freedom hurt not only the people around us, but ourselves as well.

I am by no means advocating a return to an older morality, or the Victorian ideal. There was too much wrong with the old idea—the denial of human sexuality, the double standard, the pervasive paternalism, the idealization of women, the over-control of children. Yet it is nonetheless true that in many respects the America of 1815–70 was a better society than this one. Yes, most people worked very hard; children died young; the towns and growing cities had their share of prostitution and disease; and there was always that awful institution, slavery. But the great central mass of Americans was living in a social system that was predictable, stable, and basically decent. And it was so because—despite the hypocrisy—most people felt that they had duties and obligations to other people which came before their own gratification.

In a society made up of small communities, the needs of the social group are usually clear, and plainly felt. People know that the poor and helpless must be provided for, because the poor and helpless are old Mr. Swenson in the corner house who is crippled with arthritis, and the six children of Emma Brown whose husband is an alcoholic. In small communities people realize that something must be done about the chemicals the printing plant is pouring into the river, because the fish they draw from it are beginning to taste funny. In small communities people grow up knowing that the rules and customs must be obeyed, because if Father did not see them cutting up, Aunt Agnes or Dr. Smithers will have. In small communities the people are real, and the needs of the group are apparent.

This is not the case in urban society. The poor and helpless are no longer present, because they have been shoved off into neighborhoods where those who have the means to provide charity do not see them, much less have personal feelings for them. Pollution is not caused by Riverroad Mill, but is somehow just there. Crime, drugs, the problems of the schools are newspaper stories, which touch most people in a personal way only infrequently. Today the "community" in which most people live is an abstraction, comprised of millions of people who are merely numbers.

Unfortunately, the human being was devised to respond to real people who could be seen, touched, heard; we were not created to respond viscerally to abstractions. What we must do, if we are to make any improvement in our visible decaying society, is, by means of our intelligences and imaginations, understand that we are, whether we like it or not, members of a community, or rather, sets of communities—neighborhoods surrounded by cities and towns, enclosed in counties and states and finally America. We must come to see that this America is our community, and that, as members of it, we are going to be damaged one way or another if we do not from time to time put the interests of the whole above our own concerns. A people who will not sacrifice for the common good cannot expect to have any common good. It is now abundantly clear that in large democracies a law which a substantial number of people will not obey cannot be enforced. We have seen this effect in the 1920s' laws prohibiting the sale of alcohol, the current laws against drug selling, the speeding laws, and much else. If the people will not abide by a law the government is helpless. Governments alone cannot solve our major problems. And that means that a society will only improve when the people who constitute it decide to improve it.

The American electorate for some decades has pendulated between liberalism and conservatism, however you define those two doctrines, now picking the candidates of one persuasion, now those of another. Neither the liberals nor the conservatives have been able to capture the American mind, and I think that is because neither has offered the public a consistent and well-thought-out program for the *nation as a whole*. As political scientists have said repeatedly, government in the United States today is driven by "special interest" politics. That is to say, policies and laws are worked out by compromising the demands of the most vocal groups, whether they be industry, consumer advocates, right-to-lifers, crusaders for gun control, feminists, ethnic groups, gays— the list goes on and on.

Unfortunately, this way of making law frequently sacrifices the good of the nation to the selfish needs of the competing groups. Both liberal and conservative factions have become dominated by these special-interest groups, and find themselves twisting and turning to chart courses which will keep their separate constituencies happy, rather than trying to devise broad policies for the country as a whole. The consequence is that both liberals and conservatives have become mainly sponsors for groups demanding "rights," some of which would have left the Founding Fathers with their mouths hanging open.

We have had, however, no corresponding cry for responsibilities. It is always rights: the right of the affluent to hole themselves up in their suburban fortresses without any corresponding responsibility for the central cities that produce so much of their wealth; the right of women to have babies, without any responsibility for taking care of those babies themselves; the right of non-English speakers to have their children taught in their own language, without any responsibility to see that those children study what they are taught; the right of billionaire entrepreneurs to buy up giant corporations and destroy them, without any corresponding responsibility for the welfare of their employees, suppliers, and customers; the right of young men to refuse to use birth control without any corresponding responsibility for the resulting children; the right of the arms industry to sell guns to lunatics, without any responsibility for the deaths of small children blasted off playgrounds by madmen with assault rifles; the right of doctors and lawyers to get rich from the misery of other people, without any responsibility to see that the poor have the same access to good medicine and law as the wealthy; the right of grown people in the media and the advertising industry to deceive children, without any responsibility to control the violence they show them; the right of the press to operate without any government intervention, without any corresponding responsibility to see that hard issues are pressed on their readers, instead of being jammed on page thirty-three between Dear Abby and a recipe for gooseberry fool; the right of immigrants to pour into this country without any corresponding responsibility to contribute something to the society which has given them shelter; in sum, the right of everybody and anybody to take whatever they can get without any responsibility for putting something back into the pot. The list is endless, and it includes every segment of the American society—every ethnic group, every profession, every social group.

I believe that Americans are growing tired of hearing about "rights." I am convinced that Americans are sick of going to the polls to choose among amiable opportunists who promise only to allow them to welter in cocaine and violent movies and credit card debt. I am convinced that if there arose a leader who promised to show Americans where their duty lay, he would be swept into office with astonishing majorities from all segments of the society. We have allowed ourselves to be victimized by "leaders" from all parties whose only function seems to be to find out what the electorate wants and give it to them. They are shepherds following along behind the flock, singing "Ba, Ba, Black Sheep," and keeping time with their crooks while the flock heads for the cliff. Leaders who do not have firm ideas of where the society ought to go should not present themselves for office, and they should be firmly rejected by the people when they do.

But this requires that the people take the trouble to find out what has gone wrong, and what needs to be done. It requires us to turn off our television sets, put out our joints, shut down our computer games, leave the golf course, put away our guitars; and try to find out what is going on. For in the end, in a democracy it all depends upon the people. If the people will not insist that their leaders do what needs to be done, and show a willingness to pay for it, nothing will change.

A damaged society cannot be improved by tinkering with monetary policy or in somehow "changing the system." It is critically important for us to understand that there is no such thing as a "system." What appears to be one is simply the aggregate behavior of the people who make up the society. As a consequence, a society can only be improved when those who constitute it decide to improve it. And this means making sacrifices individual by individual for the good of the whole. A government cannot legislate against the indulgent self. Only the people, acting from the springs of their own hearts, can do that. Will they? That is a question I cannot answer.

America was once more than simply a place, more than simply a nation. It was an *idea*—an idea so powerful that it inflamed the imaginations of men and women around the world, and led them everywhere to topple emperors and kings. The world no longer admires the United States. It envies our prosperity and our freedoms; but it does not admire us. Yes, immigrants continue to swarm in, but that is mainly for the abundance of things that we have. They do not come because of an idea. And Liberty weeps to see what we have done with her gift.

Notes

CHAPTER 1

1. Rudolf and Margot Wittkower, *Born Under Saturn* (New York: Norton, 1963), 166.
2. Richard Hofstadter, *Anti-Intellectualism in American Life* (New York: Vintage, 1963), 82.
3. Norman H. Clark, *Deliver Us from Evil* (New York: Norton, 1976), 21.
4. Mark Lender and James Kirby Martin, *Drinking in America* (New York: The Free Press, 1982), 14.
5. Jack Larkin, *The Reshaping of Everyday Life* (New York: Harper and Row, 1988), 281.
6. John D'Emilio and Estelle B. Freedman, *Intimate Matters* (New York: Harper and Row, 1988), 42.
7. Daniel Scott Smith and Michael S. Hinds, "Premarital Pregnancy in America 1640–1971: An Overview Interpretation," *Journal of Interdisciplinary History 4* (Spring 1975): 537.
8. Jack Larkin, *Reshaping of Everyday Life,* 283.
9. Jack Larkin, ibid., has a good discussion of the folkways of the period.

CHAPTER 2

1. Christopher Collier and James Lincoln Collier, *Decision in Philadelphia* (New York: Random House, 1986), 23.
2. Ibid., 11–13.
3. Daniel Walker Howe, "Victorian Culture in America," 3–28, in Daniel Walker Howe, ed., *Victorian America* (Philadelphia: Univ. of Pennsylvania Press, 1976), 4.
4. Stowe Persons, *The Decline of American Gentility* (New York: Columbia Univ. Press, 1973), 4.
5. David D. Hall, in Howe, ed., *Victorian America,* 87.
6. See in particular Stuart M. Blumin, *The Emergence of the Middle Class* (Cam-

265

bridge: Cambridge Univ. Press, 1989); Mary P. Ryan, *Cradle of the Middle Class* (Cambridge: Cambridge Univ. Press, 1981); Paul E. Johnson, *A Shopmaker's Millennium* (New York: Hill and Wang, 1978).

7. Larkin, *The Reshaping of Everyday Life*, chaps. 3 and 4 passim. See also Susan E. Hirsch, *Roots of the American Working Class* (Philadelphia: Univ. of Pennsylvania Press, 1978), 5.

8. Ryan, *Cradle of the Middle Class*, 75–80.

9. It is a statistical impossibility for the reduction of premarital pregnancy rates from 30% to 10% to have been produced solely by the urban middle class.

10. Hall, in Howe, ed., *Victorian America*, 90.

11. Persons, *The Decline of American Gentility*, 66.

12. Robin Gilmour, *The Idea of the Gentleman in the Victorian Novel* (London: Allen and Unwin, 1981), 29.

13. Persons, *Decline of American Gentility*, 3.

14. John R. Reed, *Victorian Conventions* (Athens: Ohio University Press, 1975), 7.

15. Quoted, ibid., 7.

16. Steven Mintz and Susan Kellogg, *Domestic Revolutions: A Social History of Family Life* (New York: The Free Press, 1988), 54.

17. Ibid., 46.

18. Persons, *Decline of American Gentility*, 88.

19. Ryan, *Cradle of the Middle Class*, 99–101.

20. Larkin, *The Reshaping of Everyday Life*, 52.

21. Mintz and Kellogg, *Domestic Relations*, 48.

22. Hofstadter, *Anti-Intellectualism in American Life*, 82.

23. Ibid., 100.

24. Reed, *Victorian Conventions*, 20.

25. Ibid., 17.

26. Diane Kelder, *The Great Book of French Impressionism* (New York: Abbeville Press, 1980), 29, 74.

27. Henry-Russell Hitchcock, *Early Victorian Architecture in Britain* (New York: Da Capo, 1976), 13.

28. Lawrence W. Levine, *Highbrow Lowbrow* (Cambridge: Harvard Univ. Press, 1988), 85–168.

29. Ibid., 7.

30. Ann Douglas, *The Feminization of American Culture* (New York: Anchor Press, 1977), 36.

31. Larkin, *The Reshaping of Everyday Life*, 300.

32. Ibid., 132.

33. Persons, *Decline of American Gentility*, 66.

34. Johnson, *A Shopkeeper's Millennium*, 55.

35. Ibid., 57.

36. Joseph R. Gusfield, "Prohibition: The Impact of Political Utopianism," in John Braeman, Robert H. Bremner, and David Brody, eds., *Change and Continuity in Twentieth Century America: The 1920s* (Athens: Ohio State University Press, 1968), 272–73.

37. Clark, *Deliver Us from Evil*, 42–44.

38. Herbert Asbury, *The Great Illusion* (Garden City: Doubleday, 1950), 156.

39. Mintz and Kellogg, *Domestic Revolutions*, 55.
40. Ian Watt, *The Rise of the Novel* (Berkeley: Univ. of California Press, 1957), 160–61.
41. Ibid.
42. Ibid., 163.
43. D'Emilio and Freedman, *Intimate Matters*, 159.
44. James Collier, *The Hypocritical American* (Indianapolis: Bobbs-Merrill, 1964), passim.
45. Ibid., 189.
46. Smith and Hinds, *Premarital Pregnancy*, 537.
47. D'Emilo and Freedman, *Intimate Matters*, 180.
48. Catherine Fennelly, *The Garb of Country New Englanders 1790–1840* (Sturbridge, Mass.: Old Sturbridge Village, 1966), 4–5.
49. Valerie Steele, *Fashion and Eroticism* (New York: Oxford Univ. Press, 1985), 51.
50. Ibid., 51–52.
51. Ibid. c. 144.
52. Elaine Tyler May, *Great Expectations: Marriage and Divorce in Post-Victorian America* (Chicago: Univ. of Chicago Press, 1980), 34–35.
53. Collier, *The Hypocritical American*, 166.
54. May, *Great Expectations*, 20.
55. Richard D. Brown, in Howe, ed., *Victorian America*, 43.
56. Quoted in Persons, *Decline of American Gentility*, 153.

CHAPTER 3

1. This description of eighteenth-century America is taken, in the main, from Collier and Collier, *Decision in Philadelphia*, 14–24.
2. Robert Redfield, *The Little Community* (Chicago: Univ. of Chicago Press, 1967), 3.
3. Ibid.
4. Ryan, *Cradle of the Middle Class*, 30–31.
5. Collier and Collier, *Decision in Philadelphia*, 19.
6. Ryan, *Cradle of the Middle Class*, 42.
7. Ibid., 43.
8. Maldwyn Allen Jones, *American Immigration* (Chicago: Univ. of Chicago Press, 1960), 63.
9. Barbara M. Tucker, *Samuel Slater and the Origins of the American Textile Industry 1790–1860* (Ithaca: Cornell Univ. Press, 1984), passim.
10. Alan Kraut, *The Huddled Masses: The Immigrant in American Society 1880–1921* (Arlington Heights, Ill.: Harlan Davidson, 1982), 64.
11. The Immigration Commission, *Abstract of the Statistical Review of Immigration to the United States 1820 to 1910* (Washington, D.C.: Government Printing Office, 1911), 8.
12. John Higham, *Strangers in the Land* (New Brunswick, N.J.: Rutgers Univ. Press, 1988), 196.
13. Thomas J. Archdeacon, *Becoming American: An Ethnic History* (New York: The Free Press, 1983), 113–18.
14. Higham, *Strangers in the Land*, 17.

15. Jones, *American Immigration,* 190–91.
16. Larkin, *Reshaping of Everyday Life,* 149–50.
17. Lawrence Levine, *Highbrow Lowbrow,* 175.
18. Ibid., 172.
19. Higham, *Strangers in the Land,* 44.
20. Stowe Persons, "The Americanization of the Immigrant," in Bowers, *Foreign Influences in American Life* (Princeton: Princeton Univ. Press, 1944), 41.
21. *Abstract of the Statistical Review of Immigration to the United States 1820 to 1910,* 50.
22. Archdeacon, *Becoming American,* 139.
23. Ibid.
24. Kraut, *Huddled Masses,* 78.
25. Ibid.
26. Archdeacon, *Becoming American,* 139.
27. Kraut, *Huddled Masses,* 113.
28. Ibid., 118.
29. John Bodnar, *The Transplanted: A History of Immigrants in Urban America* (Bloomington: Indiana Univ. Press, 1985), 189.
30. Kraut, *Huddled Masses,* 125–26.
31. Ibid., 149.
32. Ibid., 22.
33. Women's Department of the National Civic Federation, New York–New Jersey Section, *Report of the Committee on Immigration* (New York: 1916), 3.
34. John Higham, *Send These to Me* (Baltimore: Johns Hopkins Univ. Press, 1984), 25.
35. Kraut, *Huddled Masses,* 104.
36. Ibid.
37. Ibid., 102.
38. Ibid., 138; Bodnar, *The Transplanted,* 194.
39. Kraut, *Huddled Masses,* 80.
40. Archdeacon, *Becoming American,* 118.
41. Higham, *Strangers in the Land,* 65.
42. Kraut, *Huddled Masses,* 25.
43. Ibid., 65.
44. Frederic S. Conningham, ed., *Currier and Ives Prints* (New York: Crown, 1983), xiii. The print-makers' 350 scenes of farm and country life were among their most popular works.
45. The American Society of Composers, Authors and Publishers, *ASCAP Hit Tunes* (undated), 1–3.
46. Egal Feldman, "Prostitution, the Alien Woman and the Progressive Imagination," *American Quarterly* 19 (Summer 1967): 194.
47. Richard D. Brown, "Modernization in the Victorian Climax," in Howe, ed., *Victorian America,* 37.
48. Ibid., 43.
49. Higham, *Strangers in the Land,* 25.
50. Lawrence B. Davis, *Immigrants, Baptists, and the Protestant Mind in America* (Urbana: Univ. of Illinois Press, 1973), 53.
51. Kraut, *Huddled Masses,* 133.
52. Davis, *Immigrants,* 54.

53. Lender and Martin, *Drinking in America*, 96.
54. Albert D. Klassen, Colin J. Williams, and Eugene E. Levitt, *Sex and Morality in the U.S.* (Middletown, Conn.: Wesleyan University Press, 1989), 41.
55. Paul W. McBridge, *Culture Clash—Immigrants and Reformers 1880–1920* (San Francisco: R. and E. Research Associates, 1975), 15–16.
56. Kraut, *Huddled Masses*, 126.

CHAPTER 4

1. Ray Ginger, *Age of Excess* (New York: Macmillan, 1965), 39.
2. Kraut, *Huddled Masses*, 86.
3. Higham, *Send These to Me*, 23.
4. Quoted in Norman F. Cantor and Michael S. Wertham, *The History of Popular Culture Since 1815* (New York: Macmillan, 1968), 115.
5. E. Digby Baltzell, ed.: *The Search for Community in Modern America* (New York: Harper and Row, 1968), 6.
6. The Immigration Commission, *Abstract of the Statistical Review of Immigration to the United States 1820 to 1910*, 9.
7. Robert S. Lynd and Helen Merrell Lynd, *Middletown* (New York: Harcourt Brace Jovanovich, 1929), 35.
8. Robert S. Lynd and Helen Merrell Lynd, *Middletown in Transition* (New York: Harcourt Brace Jovanovich, 1937), 453.
9. Kraut, *Huddled Masses*, 67.
10. Collier and Collier, *Decision in Philadelphia*, 17.
11. Hirsch, *Roots of the American Working Class*, 12–13.
12. Ibid., 7; Blumin, *Emergence of the Middle Class*.
13. Richard Hofstadter, quoted by E. Digby Baltzell, *The Search for Community in Modern America*, 6.
14. Theodore Caplow, Howard M. Bahr, Bruce A. Chadwick, Reuben Hill, Margaret Holmes Williamson, *Middletown Families* (Minneapolis: Univ. of Minnesota Press, 1982), 10.
15. Edward Pessen, *Riches, Class and Power Before the Civil War* (Lexington, Mass.: D. C. Heath, 1973), 22–23.
16. C. Wright Mills, *White Collar* (New York: Oxford Univ. Press, 1951), 5.
17. Blumin, *Emergence of the Middle Class*, 271.
18. Richard F. Curtis and Elton F. Jackson, *Inequality in American Communities* (New York: Academic Press, 1977), 88–89. Their survey data show men placing themselves either in "Working class" or one of several middle-class categories.
19. Bodnar, *The Transplanted*, 170.
20. May, *Great Expectations*, 50.
21. Lynd and Lynd, *Middletown*, 23–24.
22. Ibid., 22–23.
23. Caplow et al., *Middletown Families*, 88.
24. Archdeacon, *Becoming American*, 135.
25. Robert A. Woods and Albert J. Kennedy, *Young Working Girls: A Summary of Evidence from Two Thousand Social Workers* (Boston: Houghton Mifflin, 1913), 68.

26. Arthur S. Link and Richard L. McCormick, *Progressivism* (Arlington Heights, Ill.: Harlan Davidson, 1983), 77.

27. Kraut, *Huddled Masses*, 72.

28. Quoted, ibid., 70.

29. Jacob A. Riis, *How the Other Half Lives*, passim.

30. Mintz and Kellogg, *Domestic Revolutions*, 84.

31. Bodnar, *The Transplanted*, 57.

32. Ibid., 76.

33. Ibid., 72.

34. Lynd and Lynd, *Middletown*, 135.

35. Archdeacon, *Becoming American*, 112.

36. Ibid., 135.

37. The Immigration Commission, *Abstract of the Statistical Review of Immigration to the United States 1820 to 1910*, 49.

38. *Historical Statistics of the United States*, 12.

39. Mintz and Kellogg, *Domestic Revolutions*, 5.

40. Ibid., 13.

41. Kurt H. Wolff, ed., *The Sociology of Georg Simmel* (New York: The Free Press, 1950), 416.

42. Lynd and Lynd, *Middletown*, 60.

43. Bodnar, *The Transplanted*, 175.

44. Lynd and Lynd, *Middletown*, 119.

45. Bodnar, *The Transplanted*, 211.

46. Ibid., 65.

47. Lynd and Lynd, *Middletown*, 66–67.

48. Lynd and Lynd, *Middletown in Transition*, 453.

49, Larkin, *Reshaping of Everyday Life*, 271.

50. Ibid., 266–70.

51. Howard P. Chudacoff, *The Evolution of American Urban Society* (Englewood Cliffs, N.J.: Prentice-Hall, 1975), 49.

CHAPTER 5

1. *Report and Recommendations of Morals Efficiency Commission* (Pittsburgh, 1913), 21.

2. Minneapolis Vice Commission, *Report of the Vice Commission of Minneapolis* (Minneapolis, 1911), 45.

3. Howard A. Kelly, *The Double Shame of Baltimore: Her Unpublished Vice Report and Her Utter Indifference* (Baltimore, undated), 15.

4. D'Emilio and Freedman, *Intimate Matters*, 148.

5. Ibid.

6. St. Louis Health Department, *Prostitution in Relation to the Public Health* (St. Louis, 1873), 8–10.

7. Ruth Rosen, *The Lost Sisterhood: Prostitution in America, 1900–1918* (Baltimore: Johns Hopkins Univ. Press, 1982), 54.

8. St. Louis Health Department, *Prostitution in Relation to the Public Health*, 11–12.

9. Charles Henry Parkhurst, *Our Fight with Tammany* (New York: Scribner's, 1895), 96.
10. *Report of the Vice Commission of Minneapolis*, 23; Report of the Little Rock Vice Commission (Little Rock, 1913), 9–10.
11. Walter C. Reckless, *Vice in Chicago* (Chicago: Univ. of Chicago Press, 1933), 271.
12. Rosen, *Lost Sisterhood*, 42.
13. *Report of the Commission for the Investigation of the White Slave Traffic, Called February, 1914* (no other data given), 12. This was actually the Massachusetts Vice Commission's report.
14. David Lawrence, *Washington-Cleanest Capital in the World* (New York: American Social Hygiene Association, 1917), 315.
15. Parkhurst, *Our Fight with Tammany*, passim.
16. Robert C. Harland, *The Vice Bondage of a Great City* (Chicago: The Young People's Civic League, 1912), 87.
17. Joseph Mayer, *The Regulation of Commercial Vice* (New York: Klebold Press, 1922), 11; *Report of the Vice Commission, Louisville, Kentucky* (Louisville, 1915), 18.
18. D'Emilio and Freedman, *Intimate Matters*, 50.
19. Al Rose, *Storyville, New Orleans* (University: Univ. of Alabama Press, 1974), 7.
20. D'Emilio and Freedman, *Intimate Matters*, 51.
21. Ibid., 50.
22. Mayer, *Regulation of Commercial Vice*, 6.
23. Rose, *Storyville*, ix, 167.
24. *Report of the Vice Commission of Minneapolis*, 23.
25. *Report of the Little Rock Vice Commission*, 9.
26. Emmett Dedmon, *Fabulous Chicago*, 252–55.
27. Ibid., 253.
28. Ibid.
29. Ibid.
30. Rose, *Storyville*, 81.
31. Ibid., 82.
32. *Report of the Hartford Vice Commission, Hartford, Conn.* (July 1913), 23–24.
33. Rose, *Storyville*, 159.
34. *Report of the Little Rock Vice Commission*, 9.
35. The Vice Commission of Philadelphia, *A Report on Existing Conditions* (Philadelphia, 1913), 16–17.
36. John Szarkowski, ed., *E. J. Bellocq: Storyville Portraits* (New York: The Museum of Modern Art, 1970), passim.
37. *Report of the Hartford Vice Commission*, 23–24; *A Report on Existing Conditions* (Philadelphia), 17; *Report and Recommendations of Morals Efficiency Commission* (Pittsburgh), 32–33; *Report of the Commission for the Investigation of the White Slave Traffic* (Massachusetts), 35.
38. *A Report of Existing Conditions* (Philadelphia), 17.
39. St. Louis Health Department, *Prostitution in Relation to the Public Health*, 11.
40. *Report of the Commission for the Investigation of the White Slave Traffic* (Massachusetts), 35.

41. My estimates based on ages given in Massachusetts, Pittsburgh, and Hartford reports.
42. Quoted in Lloyd Wendt and Herman Kogan, *Lords of the Levee* (Indianapolis: Bobbs-Merrill, 1943), 294–95.
43. *Report of the Hartford Vice Commission,* 70.
44. *Report of the Commission for the Investigation of the White Slave Traffic* (Massachusetts), 30.
45. Rosen, *Lost Sisterhood,* 22.
46. Paul A. Mertz, *Mental Deficiency of Prostitutes* (Washington, D.C.: 1919), 6.
47. Ibid., 8–9.
48. *Report and Recommendations of Morals Efficiency Commission,* (Pittsburgh), 30.
49. *Report of the Commission for the Investigvation of the White Slave Trade* (Massachusetts), 42.
50. *Report and Recommendations of Morals Efficiency Commission* (Pittsburg), 32.
51. Woods and Kennedy, *Young Working Girls,* 56.
52. Ibid., 41.
53. *A Report on Existing Conditions* (Philadelphia), 117.
54. Ibid., 85–89.
55. *Report of the Commission for the Investigation of the White Slave Trade* (Massachusetts), 20.
56. Woods and Kennedy, *Young Working Girls,* 106.
57. *Report of the Vice Commission of Minneapolis,* 76.
58. *Report of the Hartford Vice Commission,* 70–73.
59. Estimates from figures given in the Pittsburgh, Hartford, Massachusetts, and Philadelphia vice commission reports.
60. Philadelphia report, 17; Massachusetts report, 38; Hartford report, 45.
61. Egal Feldman, "Prostitution, the Alien Woman and the Progressive Imagination," 199.
62. D'Emilio and Freedman, *Intimate Matters,* 208.
63. Reckless, *Vice in Chicago,* 32–34.
64. Feldman, "Prostitution, the Alien Woman . . . ," 199–202.
65. *Report of the Commission for the Investigation of the White Slave Trade* (Massachusetts), 21.
66. *A Report on Existing Conditions* (Philadelphia), 16.
67. *Prostitution in Relation to the Public Health* (St. Louis), 10.
68. The Vice Commission of Chicago, *The Social Evil in Chicago* (Chicago, 1911), 98.
69. *Report of the Little Rock Vice Commission,* 9.
70. Willie the Lion Smith with George Hoefer, *Music on My Mind* (New York: Da Capo, 1975), 28.
71. Al Rose, *Storyville,* 85.
72. Ibid., 149–50.
73. Rosen, *Lost Sisterhood,* 88.
74. *A Report on Existing Conditions* (Philadelphia), 22–23.
75. Ibid., 81.
76. Ibid., 82.
77. Finis Farr, *Chicago* (New Rochelle, N.Y.: Arlington House, 1973), 307.
78. *Report of the Hartford Commission,* 33.
79. *A Report on Existing Conditions* (Philadelphia), 21.

80. *Report of the Vice Commission of Minneapolis*, 60.
81. Alfred C. Kinsey, Wardell B. Pomeroy, and Clyde E. Martin, *Sexual Behavior in the Human Male* (Philadelphia: W. B. Saunders, 1948), 351.
82. Alfred C. Kinsey, Wardell B. Pomeroy, Clyde E. Martin, Paul H. Gebhard, *Sexual Behavior in the Human Female* (Philadelphia: W. B. Saunders, 1953), 297–99.
83. Kinsey, Pomeroy, and Martin, *Sexual Behavior in the Human Male*, 410–13.
84. *Report and Recommendations of Morals Efficiency Commission*, (Pittsburgh), 33.
85. Women's League for Good Government, Elmira, N.Y., *Vice Conditions in Elmira* (Elmira, 1913), 13–19.
86. Parkhurst, *Our Fight with Tammany*, 11.
87. *Report of the Commission for the Investigation of the White Slave Trade* (Massachusetts), 24.
88. David Lawrence, *Washington*, 315.
89. Asbury, *The Barbary Coast*, 286.
90. Tom Stoddard, *Jazz on the Barbary Coast* (Chigwell, Essex, Eng.: Storyville Publications, 1982), 64.
91. Mayer, *Regulation of Commercial Vice*, 11.
92. *Report of the Commission for the Investigation of the White Slave Trade* (Massachusetts), 11–12.

CHAPTER 6

1. Clark, *Deliver Us from Evil*, 45.
2. Roy Rosenzweig, *Eight Hours for What We Will* (Cambridge: Cambridge Univ. Press, 1983), 57.
3. Clark, *Deliver Us from Evil*, 4.
4. Ibid., 9.
5. Ibid., 93.
6. Ibid., 116.
7. Ibid., 97.
8. Ibid., 107.
9. Ibid., 9.
10. Joseph R. Gusfield, in Braemen et al., *Change and Continuity in Twentieth Century America*, 291.
11. Robert Wiebe, *The Search for Order* (New York: Hill and Wang, 1967), 50–51; Maldwyn Allen Jones, *American Immigration*, 233.
12. Rosen, *Lost Sisterhood*, 54.
13. Ibid., 57.
14. David J. Pivar, *Purity Crusade: Sexual Morality and Social Control 1868–1900* (Westport: Greenwood Press, 1973), 251.
15. Ibid., 186.
16. Ibid., 207.
17. Ibid., 234–36.
18. Ibid., 185.
19. Ibid., 240.
20. *Report of the Little Rock Vice Commission*, 11.
21. Kelly, *The Double Shame of Baltimore*, 15.

22. *Report of the Commission for the Investigation of the White Slave Trade* (Massachusetts), 44.
23. Wirt W. Hallam, *The Reduction of Vice in Certain Western Cities* (1912), 33.
24. Mayer, *Regulation of Commercial Vice,* 9.
25. Hofstadter, *Anti-Intellectualism in American Life,* 197.
26. Link and McCormick, *Progressivism,* 21.
27. Quoted in Ray Ginger, 299.
28. Link and McCormick, *Progressivism,* 9.
29. Ibid., 2.
30. Quoted in ibid., 58.
31. Ibid., 73.
32. Ibid., 78.
33. John Higham, *Strangers in the Land,* 176.
34. Roy Rosenzweig, *Eight Hours for What We Will,* 93–94.
35. Link and McCormick, *Progressivism,* 40.
36. Quote by Gilman M. Ostrander, in Braeman, *Change and Continuity . . . ,* 331–32.

CHAPTER 7

1. Collier and Collier, *Decision in Philadelphia,* 10.
2. H. Wiley Hitchcock, *Music in the United States: A Historical Introduction* (Englewood Cliffs, N.J.: Prentice-Hall, 1988), 92.
3. Lawrence Levine, *Highbrow Lowbrow,* 13–81.
4. A. H. Saxon, *P. T. Barnum: The Legend and the Man* (New York: Columbia Univ. Press, 1989), 55.
5. Ibid., 90.
6. Johnson, *A Shopkeeper's Millennium,* 54.
7. Neil Harris, *Humbug* (Boston: Little, Brown, 1973), 237–38.
8. Don B. Wilmeth, *Variety Entertainment and Outdoor Amusement* (Westport, Conn.: Greenwood Press, 1982), 199.
9. Saxon, *P. T. Barnum,* 92.
10. Ibid., 70.
11. Ibid., 107–8.
12. Harris, *Humbug,* 239.
13. Robert C. Toll, *Blacking Up* (New York: Oxford Univ. Press, 1974), 26–28.
14. Wilmeth, *Variety Entertainment,* 118.
15. Toll, *Blacking Up,* 200.
16. John Richardson, *Manet* (Oxford: Phaidon Press, 1982), opposite plate 37.
17. Ibid., plate 38.
18. Kathy Peiss, *Cheap Amusements: Working Women and Leisure in Turn-of-the-Century New York* (Philadelphia: Temple Univ. Press, 1986), 142.
19. Joe Laurie, Jr., *Vaudeville: From the Honky-Tonks to the Palace* (New York: Holt, 1953), 14–15.
21. Ibid.
22. Wilmeth, *Variety Entertainment,* 131.
22. Claudia D. Johnson, in Howe, ed., *Victorian America,* 113.
23. Ibid., 116.

24. Douglas Gilbert, *American Vaudeville* (New York: Whittlesey House, 1940), 113.
25. Ibid., 243.
26. Ibid., 160.
27. Ibid.
28. Ibid., 58–59.
29. Wilmeth, *Variety Entertainment*, 133.
30. Gilbert, *American Vaudeville*, 206.
31. Ibid., 214.
32. Ibid., 201.
33. Wilmeth, *Variety Entertainment*, 134.
34. Peiss, *Cheap Amusements*, 144.
35. John E. Pfeiffer, *The Creative Explosion* (New York: Harper and Row, 1982), 180–81.
36. Hitchcock, *Music in the U.S.*, 2–15.
37. Ibid., chap. 5, pp. 96–129 passim.
38. Charles Hamm, *Yesterdays: Popular Song in America* (New York: Norton, 1983; first published 1979), 285.
39. Ibid.
40. Ibid., 286.
41. Ibid., 288.
42. Ibid., 289.
43. Ibid., 290.
44. Ibid., 296–97.
45. H. Wiley Hitchcock and Stanley Sadie, eds., *The New Grove Dictionary of American Music*, Vol. 3: *Piano*, Cynthia Adams Hoover (London: Macmillan, 1986), 559–62.
46. Ibid., 560.
47. Ibid., 561.
48. Hamm, *Yesterdays*, 324–25.
49. Ibid., 325.
50. Ibid., 315.
51. Alec Wilder, *American Popular Song* (New York: Oxford Univ. Press, 1972), 93.
52. Hamm, *Yesterdays*, 291.
53. James Lincoln Collier, *The Making of Jazz* (Boston: Houghton Mifflin, 1978), 16–71.
54. Ibid.
55. See Dena J. Epstein, *Singful Tunes and Spirituals* (Urbana: Univ. of Illinois Press, 1977), 241–51.
56. Eileen Southern, *The Music of Black Americans* (2nd ed., New York: Norton, 1983), 249–51.
57. Thomas Laurence Riis, *Black Musical Theatre in New York, 1890–1915*, (doctoral dissertation, University of Michigan, 1981), passim.
58. Ibid., 124–25.
59. Given in Hitchcock, *Music in the U.S.*, 111.
60. Edward A. Berlin, *Ragtime: A Musical and Cultural History* (Berkeley: Univ. of California Press, 1980), 81.

61. The connoisseur and collector Lewis Graham, quoted in Hitchcock, *Music in the U.S.*, 121.
62. Ibid.
63. James Lincoln Collier, *The Reception of Jazz in America* (Brooklyn, N.Y.: Institute for Studies in American Music, 1988), 6.
64. Sandra Lieb, *Mother of the Blues* (Amherst: Univ. of Massachusetts Press, 1981), 3.
65. W. C. Handy, *Father of the Blues* (New York: Macmillan, 1941), 78.
66. Collier, *The Reception of Jazz in America*, 7.
67. Ibid., 12.

CHAPTER 8

1. Roland Gelatt, *The Fabulous Phonograph, 1877–1977* (New York: Collier Books, 1977), 32; John L. Fell, *A History of Films* (New York: Holt, Rinehart and Winston, 1979), 11.
2. Fell, *History of Film*, 7–11.
3. Ibid., 11.
4. Ibid., 39.
5. Ibid., 14.
6. Robert C. Allen in *The American Movie Industry*, Gorham Kindem, ed. (Carbondale: Southern Illinois Univ. Press, 1982), 12.
7. Ibid.
8. Daniel J. Czitrom, *Media and the American Mind* (Chapel Hill: Univ. of North Carolina Press, 1982), 41–42.
9. Fell, *History of Film*, 17.
10. Ibid., 19.
11. Ibid., 24.
12. Ibid., 58–59.
13. Ibid., 57.
14. Russell Merritt in *The American Film Industry*, Tino Balio, ed. (Madison: Univ. of Wisconsin Press, 1976), 60.
15. Ibid., 67.
16. Fell, *History of Film*, 99.
17. Merritt, in *American Film Industry*, 60.
18. Ibid., 73.
19. Ibid.
20. Ibid.
21. Czitrom, *Media*, 42.
22. Reuel Denney in *American Perspectives: The National Self-Image in the Twentieth Century*, Robert E. Spiller and Eric Larrabee, eds. (Cambridge, Mass.: Harvard Univ. Press, 1961), 166.
23. Merritt, in *American Film Industry*, 75.
24. Lary May, *Screening Out the Past* (New York: Oxford Univ. Press, 1980), xii.
25. Gelatt, *Fabulous Phonograph*, 29.
26. Ibid., 44–45.
27. Ibid., 60.

28. R. D. Darrell, pioneer discographer, personal communication.
29. Gelatt, *Fabulous Phonograph,* 97–99.
30. Ibid., 115.
31. Ibid.
32. Ibid., 151.
33. See issues of the *Chicago Defender* for the relevant years.
34. A collector of early record catalogues is available in the New York Public Library at Lincoln Center.
35. James Lincoln Collier, *Duke Ellington* (New York: Oxford Univ. Press, 1987), 66–71.
36. Richard D. Mandell, *Sport: A Cultural History* (New York: Columbia Univ. Press, 1984), 132–33.
37. Ibid., 161–62, 171–72.
38. Allen Guttman, *Sports Spectators* (New York: Columbia Univ. Press, 1983), 87–88.
38. Mandell, *Sport,* 153.
39. Mandell, *Sport,* 153.
40. Guttman, *Sports Spectators,* 64.
41. Jesse Frederick Steiner, *Americans at Play* (New York: McGraw-Hill, 1933), 1–5.
42. Ibid., 8.
43. Mandell, *Sport,* 180–84.
48. Ibid., 184.
45. Guttman, *Sports,* 89–90.
46. Ibid., 98.
47. Frederic L. Paxon in *The Sporting Image: Readings in American Sport History,* Paul J. Zingg, ed. (Lanham, Md.: University Press of America, 1988), 47.
48. Guttman, *Sports,* 93.
49. Paxon, *Sporting Image,* 47.
50. Ibid.
51. Mandell, *Sport,* 184.
52. Ibid., 188.
53. Paxon, *Sporting Image,* 58.
54. Guttman, *Sports,* 95.
55. John R. Betts in *Sporting Image,* 177.
56. Ibid.
57. Steven A. Riess in *Sporting Image,* 248.
58. Mandell, *Sport,* 190.
59. Riess in *Sporting Image,* 248.
60. Ibid., 252.
61. Mandell, *Sport,* 184.
62. Betts, in *Sporting Image,* 182.
63. Ibid., 172.
64. Women's Department of the National Civic Federation, New York-New Jersey Section: *Report of the Committee on Immigration,* 22.
65. Steiner, *Americans at Play,* 167.
66. Mandell, *Sport,* 188–89.

CHAPTER 9

1. A. J. Ayer, *Philosophy in the Twentieth Century* (New York: Vintage, 1984), 75.
2. Ibid., 78.
3. Ibid., 82.
4. Ibid., 79.
5. Hofstadter, *Anti-Intellectualism in American Life,* 373.
6. Ibid., 381.
7. Ibid., 364.
8. Ibid., 348.
9. Sigmund Freud, *The Basic Writings of Sigmund Freud* (New York: The Modern Library, 1938), passim.
10. Sigmund Freud, *Civilization and Its Discontents* (New York: Norton, 1961), passim.
11. I. P. Pavlov, *Conditioned Reflexes* (New York: Dover, 1960; first published 1927), 1–42.
12. Henry F. May, *The End of American Innocence*, (New York: Knopf, 1959), 176.
13. Daniel Walker Howe, "1919: Prelude to Normalcy," 3–31, in Howe, ed., *Victorian America,* 18.
14. John R. Reed, *Victorian Conventions* (Athens: Ohio Univ. Press, 1975), 20.
15. Henry F. May, *The End of American Innocence,* 159.
16. *Dreiser, Mencken Letters,* Vol. I, p. 59.
17. Ibid., Vol. II, p. 740.
18. Ibid., Vol. I, pp. 57–58.
19. Julian Symons, *Makers of the New* (New York: Random House, 1987), 20–25.
20. Ibid., 53.
21. May, *End of American Innocence,* 265.
22. Symons, *Makers of the New,* 139.
23. Bennard B. Perlman, *Painters of the Ashcan School: The Immortal Eight* (New York: Dover, 1979), passim.
24. Herbert, *Impressionism,* 28.
25. Perlman, *Painters of the Ashcan School,* 52.
26. Ibid. 90.
27. Ibid., 87–88.
28. Ibid., 90.
29. Ibid., 97.
30. Ibid., 118.
31. Ibid., 178.
32. Ibid., 203.
33. Ibid., 207.
34. Robert Shiff, "The End of Impressionism," in Charles S. Moffett, ed., *The New Painting: Impressionism 1874–1886* (Geneva: Richard Burton, 1986), 83.
35. Martin Green, *The Armorg Show and the Paterson Strike Pageant* (New York: Scribner, 1988), 181.
36. Ibid., 179–80.

37. Perlman, *Painters of the Ashcan School,* 196.
38. Beaumont Newhall, *The History of Photography* (New York: The Museum of Modern Art, 1949), 138.
39. Eric Salzman, *Twentieth Century Music* (Englewood Cliffs, N.J.: Prentice-Hall, 1988), 46.
40. Ibid., 4.
41. Ibid., 29.
42. Hitchcock, *Music in the United States,* 146.
43. Salzman, *Twentieth Century Music,* 172.
44. Ibid., 134.
45. Hitchcock, *Music in the United States,* 149.
46. Ibid.
47. Ibid., 170.
48. Ibid., 150.
49. Ibid., 152.
50. Salzman, *Twentieth Century Music,* 21–22.
51. Elizabeth Kendall, *Where She Danced* (Berkeley: Univ. of California Press, 1979), chap. 3, pp. 54–69 passim.
52. Ibid., 10.
53. Ibid., 86.
54. Nancy Cott, *The Grounding of Modern Feminism* (New Haven: Yale Univ. Press, 1987), 18.
55. Ibid., 13.
56. Ibid., 37.
57. Ibid., 36.
58. Ibid., 35.
59. Valerie Steele, *Fashion and Eroticism* (New York: Oxford Univ. Press, 1985), 6.
60. Ibid., 213.
61. Quoted in *Fashion and Eroticism,* 227.
62. Quoted, ibid., 226.
63. Higham, *Strangers in the Land,* 30–31.
64. Hofstadter, Miller, and Aaron, *The American Republic,* 697.
65. Ray Ginger, 299.
66. Hofstadter, Miller, and Aaron, *The American Republic,* 697.
67. Martin Green, *The Armory Show and the Paterson Strike Pageant,* 89–98.
68. T.J. Jackson Lears, *No Place of Grace* (New York: Pantheon, 1981), xiv.
69. Henry F. May, *End of American Innocence,* 221.
70. Martin Green, *The Armory Show . . . ,* 97.
71. May, *End of American Innocence.*
72. Perlman, *Painters of the Ashcan School,* 195.
73. Rebecca Zurier, *Art for the Masses* (Philadelphia: Temple Univ. Press, 1988), 37.
74. Ibid., 38.
75. Ibid., 66.
76. Ibid., 38.
77. May, *End of American Innocence,* 106.
78. Green, *The Armory Show . . . ,* 285.

CHAPTER 10

1. Lary May, *Screening Out the Past,* 27.
2. Larkin, *Reshaping of Everyday Life,* 169.
3. Caroline F. Ware, *Greenwich Village, 1920–30* (Boston: Houghton Mifflin, 1935), 238.
4. Ibid.
5. James L. Hedrick, *Smoking, Tobacco, and Health* (Resource Management Corp., L968. Prepared for the United States Department of Health, Education and Welfare).
6. Ware, *Greenwich Village,* 240.
7. Clark, *Deliver Us from Evil,* 146.
8. Asbury, *The Grand Illusion,* 121–22.
9. Gusfield, in Braeman et al., *Change and Continuity . . . ,* 285.
10. Ibid., 288. See Perlman, *Painters of the Ashcan School,* 57, 101, and passim; Lewis A Erenberg, *Stepping Out,* 114–19.
11. Gusfield, in Braeman et al., *Change and Continuity . . . ,* 288.
12. Dedmon, *Fabulous Chicago,* 120.
13. Larkin, *Reshaping of Everyday Life,* 243.
14. Marshall and Jean Stearns, *Jazz Dance* (New York: Macmillan, 1968).
15. Robert L. Herbert, *Impressionism* (New Haven: Yale Univ. Press, 1988), 136.
16. Ibid., 131.
17. T. DeWitt Talmage, *Night Side of City Life* (Chicago: J. Fairbanks, 1878), 46.
18. A good collection of such pictures is available in the picture division of the Mid-Manhattan Library in New York City.
19. Erenberg, *Stepping Out,* 155–70.
20. *New York Times,* June 25, 1922, III, 8.
21. Asbury, *The Barbary Coast,* 287.
22. Henry O. Osgood, *So This Is Jazz* (Boston: Little, Brown, 1926), 89–90.
23. Irene Castle, *Castles in the Air* (Garden City: Doubleday, 1958), 86–87.
24. Ibid., 85–86.
25. Peiss, *Cheap Amusements,* 84.
26. Ibid., 102.
27. *New York Times,* Jan. 31, 1909, Pt. 2, 8:1.
28. Ibid., March 7, 1910, 18:5.
29. The Vice Commission of Philadelphia, *A Report on Existing Conditions,* 73–77.
30. Paul G. Cressey, *The Taxi-Dance Hall* (New York: Greenwood Press, 1932), 25.
31. Ibid., passim.
32. Rosen, *Lost Sisterhood,* 159.
33. Erenberg, *Stepping Out,* 150–55.
34. Ibid., 115.
35. Fell, *A History of Films,* 82.
36. Erenberg, *Stepping Out,* 123.
37. Ibid., 124.
38. Collier, *The Reception of Jazz in America,* 7.

39. Herbert G. Goldman, *Jolson* (New York: Oxford Univ. Press, 1988), 19; Jimmy Durante and Jack Kofoed, *Night Clubs* (New York: Knopf, 1931), 42–44; Erenberg, *Stepping Out,* 190.
40. Erenberg, *Stepping Out,* 135–36.
41. Cressey, *The Taxi-Dance Hall,* 178.
42. Erenberg, *Stepping Out,* 116.
43. Ibid., 117.
44. Personal communication, Lester A. Kirkendall, pioneer sex educator.
45. Kinsey et al., *Sexual Behavior in the Human Male,* vii.
46. Ibid., 38–40.
47. Ibid., 6.
48. Ibid., 10.
49. Ibid., 396.
50. Ibid., 397.
51. Ibid., 412.
52. Ibid., 413.
53. Kinsey et al., *Sexual Behavior in the Human Female,* 243–44.
54. Ibid., 299.
55. Ibid., 298.
56. Ibid., 300.
57. Ibid., 286.
58. Ibid., 291–92.
59. Ibid., 300.
60. Ibid., 295.
61. Kinsey et al., *Sexual Behavior in the Human Male,* 347.
62. Peiss, *Cheap Amusements,* 110.
63. Ibid., 112–13.
64. Ibid., 50.
65. Edmund Wilson, *The Twenties* (New York: Farrar, Straus and Giroux, 1975), passim.
66. D'Emilio and Freedman, *Intimate Matters,* 232.
67. Nancy Cott, *The Grounding of Modern Feminism,* 53.
68. Rebecca Zurier, *Art for the Masses* (Philadelphia: Temple Univ. Press, 1988), xv–xviii.
69. May, *The End of American Innocence,* 284.
70. James Lincoln Collier, *Benny Goodman and the Swing Era* (New York: Oxford Univ. Press, 1989), 31.
71. Wiebe, *The Search for Order,* 150–51.
72. John Chynoweth Burnham, in Braeman, et al., *Change and Continuity . . . ,* 354.
73. Ibid., 351–52.
74. Pivar, *Purity Crusade,* 3.
75. Symons, *Makers of the New,* 9.
76. Wiebe, *The Search for Order,* 208.
77. Fell, *A History of Films,* 82.
78. Lary May, *Screening Out the Past,* xii.
79. *New York Times Book Review,* Feb. 21, 1988.
80. Frances Spalding, *Vanessa Bell* (New Haven: Ticknor & Fields, 1983), 92–93.

81. Asbury, *The Great Illusion,* 121–22; Gusfield, in Braeman et. al., *Change and Continuity . . . ,* 291; Clark, *Deliver Us from Evil,* 97–105.
82. David L. Weis and Joan Jurich, "Size of Community as a Predictor of Attitudes Toward Extramarital Sexual Relations," *Journal of Marriage and the Family* 47 (Feb. 1985).
83. B. K. Singh, "Trends in Attitudes Toward Premarital Sexual Relations," *Journal of Marriage and the Family* (May 1980), 390.
84. Albert D. Klassen et al., 196, 104.
85. Ira L. Reiss, *The Social Context of Premarital Sexual Permissiveness* (New York: Holt, Rinehart and Winston, 1967), 74.
86. Malcolm Cowley, *Exile's Return* (New York: Viking, 1986; first published in 1951), 47.

CHAPTER 11

1. Edmund Wilson, *The Fifties* (New York: Farrar, Straus and Giroux, 1986), xxii.
2. Frederick Lewis Allen, *Only Yesterday* (New York: Harper and Row, 1931).
3. Lynd and Lynd, *Middletown.*
4. John T. Gueenan, *American Civilization Today: A Summary of Recent Social Trends* (New York: McGraw-Hill, 1934); T. J. Woofter, Jr., *Races and Ethnic Groups in American Life* (New York: McGraw-Hill, 1933); Sophonisba P. Breckinridge, *Women in the Twentieth Century: A Study of Their Political, Social and Economic Activities* (New York: McGraw-Hill, 1933); Jesse Frederick Steiner, *Americans at Play: Recent Trends in Recreation and Leisure Time Activities* (New York: McGraw Hill, 1933).
5. Allen, *Only Yesterday,* 29, 42, 46, 58.
6. David Burner, in Braeman et al., *Change and Continuity . . . ,* 10.
7. Quoted in Paul A. Carter, *The Twenties in America* (New York: Crowell, 1968), 13.
8. Ibid., 16.
9. Allen, *Only Yesterday,* 133–34.
10. Joel Williamson, *The Crucible of Race* (New York: Oxford Univ. Press, 1984), 117.
11. Woofter, *Races and Ethnic Groups,* 75.
12. Ibid., 75–76.
13. Ibid.
14. Mintz and Kellogg, *Domestic Revolutions,* 135.
15. Lynd and Lynd, *Middletown,* 32–72 passim.
16. Gueenan, *American Civilization Today,* 56.
17. Ibid., 58.
18. Ibid., 84.
19. May, *Great Expectations,* 62.
20. Lynd and Lynd, *Middletown,* 211–15.
21. Ibid., 211.
22. Gueenan, *American Civilization Today,* 78.
23. Lynd and Lynd, *Middletown,* 170.
24. Allen, *Only Yesterday,* 80.
25. Lynd and Lynd, *Middletown,* 169.

26. Steiner, *Americans at Play,* 149.
27. Lynd and Lynd, *Middletown,* 261.
28. Ibid., 261–62.
29. Steiner, *Americans at Play,* 61.
30. Ibid., 193.
31. Lynd and Lynd, *Middletown,* 251.
32. Charles N. Glaab, in Braeman, et al., *Change and Continuity* . . . , 406.
33. Lynd and Lynd, *Middletown,* 253.
34. Jesse Frederick Steiner, *The American Community in Action* (New York: Holt, 1928), 58.
35. Lynd and Lynd, *Middletown,* 255–56.
36. Ibid., 253.
37. Ibid., 137.
38. Ibid., 272.
39. Allen, *Only Yesterday,* 83.
40. Lynd and Lynd, *Middletown,* 258.
41. Ibid., 135.
42. Ibid., 272.
43. Steiner, *Americans at Play.*
44. Ibid., 70–71.
45. Ibid., 73.
46. Ibid., 74.
47. Ibid., 77.
48. Ibid., 106.
49. Ibid., 64.
50. Ibid., 48–49.
51. Ibid., 52–53.
52. Ibid., 53.
53. Ibid., 54.
54. Ibid.
55. Ibid., 34.
56. Ibid.
57. Ibid.
58. Ibid., 35.
59. Ibid., 25.
60. Ibid., 30.
61. Ibid., 195.
62. Ibid., 91.
63. Ibid.
64. Ibid., 86.
65. U.S. Bureau o the Census, *Statistical Abstract of the United States: 1989* (Washington, D.C., 1989), 226.
66. Fell, *A History of Films,* 104.
67. Lynd and Lynd, *Middletown,* 263–64.
68. R. D. Darrell, personal communication.
69. Gelatt, *The Fabulous Phonograph,* 212.
70. Collier, *The Reception of Jazz in America,* 15–16.
71. Thomas A. DeLong, *Pops: Paul Whiteman, King of Jazz* (Piscataway, N.J.: New Century, 1983), 59.

72. Collier, *The Reception of Jazz in America*, 17.
73. Gueenan, *American Civilization Today*, 84–87.
74. Lynd and Lynd, *Middletown*, 231–32.
75. Ibid., 230.
76. Clark, *Deliver Us from Evil;* Gusfield, in Braeman et al., *Change and Continuity.*
77. *Asbury, The Great Illusion*, 137–43.
78. Ibid., 145.
79. Ibid., 150.
80. Clark, *Deliver Us from Evil*, 147.
81. Ibid., 109.
82. Gusfield, in Braeman et al., *Change and Continuity . . . ,* 278.
83. Asbury, *The Great Illusion*, 155.
84. Ibid., 220.
85. Ibid., 320.
86. Gusfield, in Braeman et al., *Change and Continuity . . . ,* 274.
87. Clark, *Deliver Us from Evil*, 146.
88. Asbury, *The Great Illusion*, 209–10.
89. Gusfield, in Braeman, et al., *Change and Continuity . . . ,* 278.
90. Ibid., 285.
91. Author's analysis.
92. Jack J. Gottsegen, *Tobacco: A Study of Its Consumption in the United States* (New York: Pitman, 1940), 1.
93. Ibid., 5.
94. Hedrick, *Smoking, Tobacco and Health*, 3, 7.
95. Kinsey et al., *Sexual Behavior in the Human Male*, 329.
96. Ibid., 377–81, 535–37.
97. Ibid., 368.
98. Lynd and Lynd, *Middletown*, 139.
99. Both the conservative *Literary Digest*, and the more sophisticated *Vanity Fair* were examined.
100. Allen, *Only Yesterday*, 74.
101. May, *Great Expectations*, 63.
102. Allen, *Only Yesterday*, 89.
103. Kinsey et al., *Sexual Behavior in the Human Female*, 299–300.
104. Allen, *Only Yesterday*, 78.
105. Steiner, *Americans at Play*, 87.
106. Carter, *Twenties in America*, 51–52.
107. Steiner, *Americans at Play*, 12.
108. Ibid., ix.
109. May, *Great Expectations*, 84.
110. Carter, *Twenties in America*, 8.
111. Gilman M. Ostrander, in Braeman et al., *Change and Continuity . . . ,* 335.
112. Burnham, in Braeman et al., *Change and Continuity . . . ,* 351–52.
113. Clark, *Deliver Us from Evil*, 152.

CHAPTER 12

1. John Kenneth Galbraith, *The Great Crash of 1929* (Boston: Houghton Mifflin, 1961; orig. pub. 1955), 16.

2. Ellis W. Hawley, *The Great War and the Search for a Modern Order* (New York: St. Martin's Press, 1979), 180.
3. Allen, *Only Yesterday*, 255.
4. Ibid., 270.
5. Ibid., 271–75.
6. Ibid., 276–77.
7. Galbraith, *The Great Crash*, 142.
8. Hawley, *The Great War*, 58.
9. Ibid., 81.
10. Ibid., 89.
11. Galbraith, *The Great Crash*, 175.
12. Gueenan, *American Civilization Today*, 4.
13. Mintz and Kellogg, *Domestic Revolutions*, 134–35.
14. Frederick Lewis Allen, *The Big Change* (New York: Harper and Row, 1988; orig. pub. 1952), 147.
15. Allen, *Since Yesterday*, 86.
16. Allen, *The Big Change*, 148–49.
17. Allen, *Since Yesterday*, 305–7.
18. Ibid., 60.
19. Kinsey et al., *Sexual Behavior in the Human Female*, 243–44.
20. Ibid., 299.
21. Ibid., 357.
22. Ibid., 423.
23. Lynd and Lynd, *Middletown in Transition*, 170.
24. Ibid., 172.
25. Ibid., 173.
26. Ibid., 174.
27. Allen, *The Big Change*, 201.
28. Lynd and Lynd, *Middletown in Transition*, 155.
29. Clark, *Deliver Us from Evil*, 146.
30. Lender and Martin, *Drinking in America*, 197.
31. Lynd and Lynd, *Middletown in Transition*, 271–72.
32. James L. Hedrick, *Smoking, Tobacco and Health*, 7.

CHAPTER 13

1. F. Leslie Smith, *Perspective on Radio and Television* (New York: Harper and Row, 1985), 8–10.
2. Daniel J. Czitrom, *Media and the American Mind* (Chapel Hill: Univ. of North Carolina Press, 1982), 66.
3. Ibid., 68.
4. Ibid., 70.
5. Smith, *Perspective on Radio and Television*, 16.
6. Czitrom, *Media and the American Mind*, 71.
7. Ibid.
8. Ibid.
9. Erik Barnouw, *A Tower in Babel: A History of Broadcasting in the United States*, Vol. 1, *To 1933* (New York: Oxford Univ. Press, 1966), 64.
10. Ibid., 104.
11. Czitrom, *Media and the American Mind*, 71.

12. Ibid., 72.
13. Ibid.
14. James T. Maher, personal communication.
15. Czitrom, *Media and the American Mind*, 74.
16. Barnouw, *A Tower in Babel*, 96.
17. Czitrom, *Media and the American Mind*, 76.
18. Ibid.
19. Ibid.
20. Smith, *Perspective on Radio and Television*, 20.
21. Ibid.
22. Barnouw, *A Tower in Babel*, 191.
23. Ibid., 210.
24. Ibid., 166.
25. Czitrom, *Media and the American Mind*, 79.
26. Gelatt, *The Fabulous Phonograph*, 255.
27. *Variety*, Nov. 12, 1930, p. 1.
28. Barnouw, *A Tower in Babel*, 224.
29. Ibid., 225–29.
30. Ibid., 228–29.
31. Ibid., 276.
32. Ibid., 273.
33. Czitrom, *Media and the American Mind*, 85.
34. Ibid., 86.
35. Ibid., 85.
36. Ibid., 81.
37. Ibid., 82.
38. Bruce Barton, *The Man Nobody Knows* (Indianapolis: Bobs-Merrill, 1925).
39. James Lincoln Collier, *Benny Goodman and the Swing Era* (New York: Oxford Univ. Press, 1989), 104.
40. Ibid., 40.
41. Collier, *The Reception of Jazz in America*, 22.
42. Collier, *Benny Goodman and the Swing Era*, 128–36, 165–67.
43. Ibid., 186–87.
44. Ibid., 38–41.
45. Ibid., 190–94.
46. Fell, *A History of Films*, 204–7.
47. Ibid., 206.
48. Ibid., 207.
49. Ibid., 208–9.
50. Ibid., 216.
51. Ibid., 218–21.
52. Ibid., 218.

CHAPTER 14

1. Marion Hargrove, *See Here, Private Hargrove* (New York: Henry Holt, 1943).
2. Richard Tregaskis, *Guadalcanal Diary* (Redhill: W. Gardner, Darton, 1943).
3. W. L. White, *They Were Expendable* (New York: Harcourt, Brace, 1942).
4. Cornelius Ryan, *The Longest Day* (New York: Simon and Schuster, 1959).

5. William Lawrence Shirer, *The Rise and Fall of the Third Reich* (New York: Simon and Schuster, 1960).
6. James L. Stokesbury, *A Short History of World War II* (New York: William Morrow, 1980).
7. Ibid., 66.
8. Ibid., 51–53.
9. Eric Maria Remarque, *All Quiet on the Western Front* (Boston: Little, Brown, 1929); Ernest Hemingway, *A Farewell to Arms* (New York: Scribner's, 1929); John Dos Passos, *Three Soldiers* (New York: George H. Doran, 1921).
10. Allen, *The Big Change*.
11. Stokesbury, *A Short History, . . .* , 163–67.
12. Ibid., 169–71.
13. Ibid., 223–27.
14. Ibid., 352–55.
15. Ibid., 374–76.
16. Allen, *The Big Change*, 166.
17. Ibid., 169.
18. Ibid.
19. Ibid., 170.
20. Ibid.
21. George Katona, *The Mass Consumption Society* (New York: McGraw-Hill, 1964), 15.
22. Ibid.
23. Ibid., 14.
24. Ibid., 18.
25. Eric F. Goldman, *The Crucial Decade—and After: America 1945–60* (New York: Vintage no date; orig. pub. 1956), 25.
26. Ibid., 146–87, 244–447.
27. Ibid., 182.
28. Katona, *The Mass Consumption Society*, 115–16.
29. Ibid.
30. Baltzell, *The Search for Community . . .* , 9.
31. Mintz and Kellogg, *Domestic Revolution*, 193.
32. Baltzell, *The Search for Community . . .* , 9–10.
33. Ben J. Wattenberg and Richard M. Scammon, *This U.S.A.* (Garden City: Doubleday, 1965), 166.
34. U.S. Bureau of the Census, *Historical Statistics of the United States:* Part 2, pp. 385–86. From 1940 to 1950 the number of A.B. degrees conferred rose about 225%.
35. Katona, *The Mass Consumption Society*, 219–20.
36. Ibid., 233.
37. Mintz and Kellogg, *Domestic Revolutions*, 203.
38. Lender and Martin, *Drinking in America*, 197.
39. U.S. Bureau of the Census, *Historical Statistics of the United States:* Part 1, p. 49.
40. Katona, *The Mass Consumption Society*, 111.
41. Ibid., 219–20.
42. Ibid., 231–32.
43. Ibid., 12. See also Michael Harrington, *The Other America: Poverty in the United States* (New York: Macmillan, 1962).

44. Allen, *Since Yesterday,* 212.
45. Collier, *Louis Armstrong,* 332; Collier, *Duke Ellington,* 133.
46. *Television Video Almanac,* 645–46.
47. Ibid.
48. Robert T. Bower, *Television and the Public* (New York: Holt, Rinehart and Winston, 1973), 3.
49. Eric F. Goldman, *The Crucial Decade,* 191.
50. Ibid., 192.
51. Ibid., 194–95.
52. Bower, *Television and the Public,* 3.
53. *TV Facts,* 142.
54. Bower, *Television and the Public,* 3.
55. *Television Almanac,* 21A.
56. *Television Information,* Winter 1987/88.
57. Quoted in *TV Facts,* 14.
58. *TV Facts,* 14, 142.
59. George Comstock, *Television in America* (Beverly Hills: Sage Publications, 1980), 29.

CHAPTER 15

1. James Lincoln Collier, *The Hypocritical American* (Indianapolis: Bobs Merrill, 1964), 167.
2. James Jackson Kilpatrick, *The Smut Peddlers* (Garden City: Doubleday, 1960), 116–17.
3. Collier, *The Hypocritical American,* 168.
4. Ibid.
5. From 1952 to 1958 the author worked as an editor for various magazine publishers who were closely concerned with the censorship issue. Much of what follows is based on personal experience.
6. Kilpatrick, *The Smut Peddlers,* 91.
7. Alex Craig, *Suppressed Books* (Cleveland: World, 1963), 143.
8. *McCall's* 84:4, April 1957; *Coronet,* 42: 68–72, July 1957; *Reader's Digest* 72: 39–41, June 1958.
9. *Playboy,* January 1974, p. 65.
10. Ibid., 63–65.
11. Quoted in *Playboy,* January 1989.
12. During the years from 1967 to 1969 the author was deeply involved in the sex education movement, designing sex education materials for major publishers, for a period in association with the Sex Information and Education Council of the United States, under the direction of Mary Calderone. Much of what follows is based on personal experience.
13. The material on Anaheim is based on interviews with Paul Cook, Lester A. Kirkendall, Sally Williams, and others involved in the case in Spring 1967 and December 1980.
14. Interview with Sally Williams, c. December 4, 1980.
15. Interview with Peter Scales, Fall 1980.
16. Interview with Scales.

17. Tom Clark, *Jack Kerouac* (New York: Harcourt Brace Jovanovich, 1984), 58–61.
18. Ibid., passim.
19. *New York Times,* Nov. 16, 1952, Section VI, 10:3.
20. Ibid., 10.
21. Jack Kerouac, *On the Road* (New York: Signet, no date; originally published by Viking, 1957), 46.
22. *New York Times,* Nov. 16, 1952, Section VI, 10.
23. Clark, *Jack Kerouac,* passim; Barry Miles, *Ginsberg,* (New York: Simon and Schuster, 1989), passim.
24. Miles, *Ginsberg,* 51.
25. Ibid., 137.
26. Clark, *Jack Kerouac,* 55–56; Miles, *Ginsberg,* 121.
27. Miles, *Ginsberg,* 103.
28. Ibid., 73.
29. Ibid., 77–78.
30. Allen Ginsberg, *Howl and Other Poems* (San Francisco: City Lights Books, 1956), 9.
31. Miles, *Ginsberg,* 47.
32. Quoted in Richard Hofstadter, *Anti-Intellectualism in American Life,* 422.
33. Ibid., 420.
34. Clark, *Jack Kerouac,* 193.
35. Gerald Nicosia, *Memory Babe: A Critical Biography of Jack Kerouac* (New York: Grove, 1983), 697; Clark, *Jack Kerouac,* 216.
36. *New York Times,* Nov. 16, 1952, Section VI, 19.
37. Miles, *Ginsberg,* 232.
38. Ibid., 196–99.
39. *Commentary,* Dec. 1957; *The Library Journal,* June 15, 1958; *Harper's* Jan. 1958; *Newsweek,* Jan. 13, 1958; *Vogue,* Sept. 1, 1957.
40. *Time,* Sept. 1, 1957.
41. Clark, *Jack Kerouac,* 169.
42. Ibid., 173.
43. Morris Dickstein, *Gates of Eden* (New York: Penguin, 1989; orig. pub. by Basic Books, 1977), 55.
44. Nicosia, *Memory Babe,* 563.
45. Miles, *Ginsberg,* 276–81.
46. Ibid., 301.
47. Ibid., 326–27.
48. Ibid., 376.
49. Ibid., 326.
50. Dickstein, *Gates of Eden,* 6.
51. Quoted in Clark, *Jack Kerouac,* 326.
52. Ibid., 205.
53. Much of what follows is based on visits made by the author to a number of communes in the Fall 1968 and interviews with commune members at the time.
54. William H. Chafe, *The Unfinished Journey: America Since World War II* (New York: Oxford Univ. Press, 1986), 327.
55. Author's interviews.

56. Daniel Yankelovich, *The New Morality: A Profile of American Youth in the 70's* (New York McGraw-Hill, 1974), 62.
57. Interviews with Michael Metelica and members of his commune.
58. Interview with George Hurd.
59. Interviews with members of the Buckland group, who requested anonymity at the time.
60. Interview with George Hurd.
61. *New Republic,* Feb. 17, 1968.
62. Interview with George Hurd.
63. Interview with Paul Gregg.
64. Interviews with members of the Buckland group.
65. Interview with George Hurd.
66. Interview with Paul Gregg.
67. Interview with Dennis Lee.
68. Interview with George Hurd.
69. *New Republic,* Feb. 17, 1968.
70. Chafe, *The Unfinished Journey,* 333–34.
71. Ibid., 320.

CHAPTER 16

1. Winston Ehrmann, *Premarital Dating Behavior* (New York: Bantam, 1960; originally published by Henry Holt, 1959).
2. Lester A. Kirkendall, *Premarital Intercourse and Interpersonal Relationships* (New York: The Julian Press, 1961).
3. Ira L. Reiss, *The Social Context of Premarital Sexual Permissiveness* (New York: Holt, Rinehart and Winston, 1967).
4. Klassen et al., *Sex and Morality in the U.S.*
5. Reiss, *The Social Context,* 8.
6. Ibid., 11.
7. Ibid., 175.
8. Ibid., 117.
9. Ibid., 123–24.
10. Klassen et al., *Sex and Morality in the U.S.,* 267.
11. Ibid., 268.
12. Ibid., 409.
13. Ibid., 139.
14. Ibid., 410.
15. Ibid., 139.
16. Ibid., 410.
17. Ibid., 25–28.
18. Ibid., 17.
19. Yankelovich, *The New Morality,* 4.
20. The Vice Commission of Philadelphia, 143.
21. *Chicago City Directory, 1911,* 521.
22. Stanton Peele, in *The Sciences,* July/Aug. 1989, p. 21.
23. Edward M. Brecher and the Editors of Consumer Reports, *Licit and Illicit Drugs* (Boston: Little, Brown, 1972), 480.

24. W. Wayne Worick and Warren E. Schaller, *Alcohol, Tobacco and Drugs* (Englewood Cliffs, N.J.: Prentice-Hall, 1977), 114.
25. Brecher et al., *Licit and Illicit Drugs*, 480.
26. Yankelovich, *The New Morality*, 62.
27. The figures for alcohol consumption are from Lender and Martin, *Drinking in America*, 196–97.
28. *Statistical Abstract of the United States, 1989*, 61.
29. Ibid., 751.
30. *TV Facts*, 142.
31. Theodore Caplow, Howard M. Bahr, et al., *Middletown Families* (Minneapolis: Univ. of Minnesota Press, 1982), 23.
32. Ibid., 25.
33. *Statistical Abstract of the United States, 1989*, 751.
34. *Middletown Families*, 25.
35. *Statistical Abstract of the United States, 1989*, 751.
36. Lender and Martin, *Drinking in America*, 196–97.
37. *Statistical Abstract of the United States, 1989*, 115.
38. Thomas B. Vischi, Kenneth R. Jones, Ella L. Shank, and Lowell H. Lima, *The Alcohol, Drug Abuse and Mental Health National Data Book, January, 1980* (Rockville, Md.: U.S. Department of Health, Education and Welfare), 16.
39. Ibid.
40. Harvey S. Siegal, "Current Patterns of Psychoactive Drug Use" in Carl D. Chamheri, James A. Inciardi, David M. Petersen, Harvey S. Siegal, O. Z. White, eds., *Chemical Dependencies, Patterns, Costs, and Consequences* (Athens: Ohio University Press, 1987), 99.
41. Ibid., 105.
42. Jerald G. Bachman, Lloyd D. Johnston, and Patrick M. O'Malley, "Smoking, Drinking, and Drug Use Among American High School Students: Correlates and Trends, 1975–79," *American Journal of Public Health* 71 (Jan. 1981): 60.
43. Ibid., 61.
44. Ibid., 59.
45. Louis Harris, *Inside America* (New York: Vintage, 1987), 77.
46. Ira E. Robinson and Davor Jedlicka, "Change in Sexual Attitudes and Behavior of College Students from 1965 to 1980: A Research Note," *Journal of Marriage and the Family* 44 (Feb. 1982): 238–39.
47. Lura F. Henze and John W. Hudson, "Personal and Family Characteristics of Cohabiting and Noncohabiting College Students," *Journal of Marriage and the Family* 36 (Nov. 1974): 723.
48. Ibid., 724.
49. Donald W. Bower and Victor A. Christopherson, "University Student Cohabitation: A Regional Comparison of Selected Attitudes and Behavior," *Journal of Marriage and the Family* 39 (Aug. 1977): 448–50.
50. Yankelovich, *The New Morality*, 4.
51. Ibid., 24–25.
52. Arthur M. Vener and Cyrus S. Stewart, "Adolescent Sexual Behavior in Middle America Revisited: 1970–73," *Journal of Marriage and the Family* 36 (Nov. 1974): 731–32.
53. Karl King, Jack O. Balswick, and Ira E. Robinson, "The Continuing Pre-

marital Sexual Revolution Among College Females," *Journal of Marriage and the Family* 39 (Aug. 1977): 455.

54. Vener and Stewart, "Adolescent Sexual Behavior," 734.
55. Ibid.
56. Minako K. Maykovich, "Attitudes Versus Behavior in Extramarital Sexual Relations," *Journal of Marriage and the Family* 38 (Nov. 1976): 695.
57. Ibid., 696.
58. Marilyn Yalom, Wenda Brewster and Suzanne Estler, "Women of the Fifties: Their Past Sexual Experiences and Current Sexual Attitudes in the Context of Mother/Daughter Relationships," *Sex Roles* 7 (Sept. 1981): 881.
59. Ibid., 883.
60. Personal communication from Ralph Ginsberg.
61. *New York Times Book Review*, Aug. 15, 1971, p. 29; Dec. 5, 1971.
62. Ibid., July 8, 1973, p. 29.
63. Shere Hite, *The Hite Report*, (New York: Macmillan, 1976).
64. *Statistical Abstract of the United States, 1989*, 85.
65. Yankelovich, *The New Morality*, 43.

CHAPTER 17

1. *Statistical Abstract of the United States, 1989*, 424.
2. Yankelovich, *The New Morality*, 62.
3. Ibid.
4. Ibid., 93.
5. Mintz and Kellogg, *Domestic Revolutions*, 206.
6. Barbara Ehrenreich, *The Hearts of Men* (Garden City: Anchor Press, 1983), 93.
7. *New York Times Book Review*, July 17, 1960, p. 8.
8. Ibid., July 3, 1966, p. 8.
9. Ibid., Aug. 15, 1971, p. 37.
10. Ibid., July 8, 1973, p. 29.
11. Ibid., Nov. 30, 1975, p. 69.
12. Ibid., May 21, 1978, p. 48.
13. Chafe, *The Unfinished Journey*, 337.
14. Joseph Veroff, Elizabeth Douvan, Richard A. Kulka, *The Inner American: A Self-Portrait from 1957 to 1976* (New York: Basic Books, 1981), 17.
15. Ibid., 103.
16. Ibid., 114.
17. Ibid., 115.
18. Ibid., 118.
19. Ibid., 479.
20. Ibid., 529.
21. Johnston et al., *Smoking, Drinking and Drug Use . . .* , 231.
22. *New York Times*, Nov. 25, 1989, p. 10.
23. Johnston et al., *Smoking, Drinking, and Drug Use . . .* , 219.
24. Lender and Martin, *Drinking in America*, 196–97.
25. Johnston et al., *Smoking, Drinking, and Drug Use . . .* , 6.
26. Ibid., 8.
27. Ibid., 227.

28. Ibid., 231.

29. Ibid., 14.

30. U.S. Department of Health and Human Services, *Morbidity and Morality Weekly Report*, Vol. 37, No. 37 (Aug. 23, 1988): 565–68.

31. *New York Times*, June 22, 1989, A16.

32. Robin Lynn Leavitt and Martha Bauman Power, "Emotional Socialization in the Postmodern Era: Children in Day Care," *Social Psychology Quarterly* 52 (March 1989): 36.

33. *Statistical Abstract of the United States, 1989*, 85.

34. Mintz and Kellogg, *Domestic Revolutions*, 204.

35. Dale Hofmann and Martin J. Greenberg, *An Irreverent Look at Big Business in Pro Sports* (Champaign, Ill.: Leisure Press, 1989), quoted in *New York Times Book Review*, June 11, 1989, p. 44.

36. *Statistical Abstract of the United States, 1989*, 226.

37. Ibid., 221.

38. Ibid., 427.

39. Ibid.

40. *New York Times*, June 24, 1989, p. 1.

41. *Statistical Abstract of the United States, 1989*, 424.

42. Ibid., 427.

CHAPTER 18

1. Comstock, *Television in America*, 29.

2. *TV Facts*, 150.

3. Author's interview with Walter Cronkite, Dec. 1979.

4. *New York Times*, Oct. 16, 1990, C8.

5. Ibid., C1, C8.

6. Bower, *Television and the Public*, 39–40.

7. *New York Times*, Oct. 16, 1990, C8.

8. Ibid.

9. *New York Times*, Aug. 5, 1990.

10. Bower, *Television and the Public*, 55.

11. Ibid., 170–72.

12. Ibid., 2.

13. Ibid.

14. Patricia Palmer, "The Social Nature of Children's Television Viewing," in Phillip Drummond and Richard Patterson, eds., *Television and Its Audience: International Research Perspectives* (London: BFI Books, 1988, 139–53), 144.

15. Comstock, *Television in America*, 113.

16. Ibid., 109.

17. Ibid., 114.

18. Walter Goodman, in *New York Times*, Dec. 3, 1989, *Arts and Entertainment*, 33.

19. Comstock, *Television in America*, 100.

20. L. Rowell Huesmann and Riva S. Bachrach, *"Differential Effects of Television Viewing on Kibbutz and City Children,"* in Drummond and Patterson, eds., *Television and Its Audience*, 172.

CHAPTER 19

1. George Peter Murdock, *Social Structures* (New York: The Free Press, 1949), 35.

2. John Cannon and Ralph Griffiths, *Oxford Illustrated History of the British Monarchy* (New York: Oxford Univ. Press, 1988), 227.

3. Henry B. Biller and Richard S. Solomon, *Child Maltreatment and Paternal Deprivation: A Manifesto for Research, Prevention and Treatment* (Lexington, Mass.: Lexington Books, 1986), 1.

4. Ibid., passim.

5. Ibid., 69.

6. Ibid., 70.

7. Ibid., 106, 108.

8. Ibid., 153.

9. Ibid., 152.

10. Ibid.

11. Ibid., 150.

12. Ibid., 3.

13. Ibid., 74–75.

14. Ibid., 67.

15. Ibid., 104.

16. *New York Times,* July 12, 1990, C14.

17. Biller and Solomon, *Child Maltreatment . . . ,* 222.

18. Reiss, *The Social Context of Premarital Sexual Permissiveness,* 173.

19. Leavitt and Power, *Emotional Socialization . . . ,* 36.

20. Suzanne E. Landis and Jo Anne L. Earp, "Day Care Center Illness: Policy and Practice in North Carolina," *American Journal of Public Health* 78 (March 1988): 311.

21. Reported in *Science News* 136 (1989): 188.

22. Beverly F. Moore, Charles W. Snow, G. Michael Poteat, "Effects of Variant Types of Child Care Experience on the Adaptive Behavior of Kindergarten Children," *American Journal of Orthopsychiatry* 58 (April 1988): 297.

23. Marian Blum, *The Day-Care Dilemma* (Lexington, Mass.: Heath, 1983), 84.

24. Ibid., 85.

25. Moore, Snow, and Poteat, "Effects of Variant Types . . . ," passim.

26. Donna King and Carol E. MacKinnon, "Making Difficult Choices Easier: A Review of Research on Day Care and Children's Development," *Family Relations* 37 (Oct. 1988): 393.

27. Moore, Snow, and Poteat, "Effects of Variant Types . . . ," 297–98.

28. Blum, *The Day-Care Dilemma,* 85.

29. Moore, Snow, and Poteat, "Effects of Variant Types . . . ," 297–98.

30. King and MacKinnon, "Making Difficult Choices . . . ," 396.

31. Ibid.

32. *Statistical Abstract of the United States, 1989,* 424.

33. Ibid., 688–89.

34. Douglas A. Smith and G. Roger Jarjoura, "Social Structure and Criminal Victimization," *Journal of Research in Crime and Delinquency* 25 (Feb. 1988): 27–52, passim.

35. *New York Times,* July 12, 1990, B8.

36. John Kenneth Galbraith, *The Affluent Society* (Boston: Houghton Mifflin, 1958).
37. Harris, *Inside America,* 355.
38. *Statistical Abstract of the United States, 1989,* 751.
38. Teresa A. Sullivan, Elizabeth Warren, Jay Lawrence Westbrook, *As We Forgive Our Debtors* (New York: Oxford Univ. Press, 1987), 3.
39. *Statistical Abstract of the United States, 1989,* 751.
40. Ibid., 699.
41. Ibid., 656.
42. Ibid., 101.
43. *New York Times,* March 25, 1990, Sec. 4, 1.
44. Ibid., Sept. 20, 1989, D1.
45. *Statistical Abstract of the United States, 1989,* 346.
46. Ibid., 125.
47. Estimate given by Federal Prosecutor Rudolph Guiliani in a television speech, Nov. 4, 1989.
48. Harris, *Inside America,* 119.
49. Veroff et al., *The Inner American . . . ,* 479.
50. Ibid., 480.
51. Ibid., 479.
52. *Statistical Abstract of the United States, 1989,* 48.
53. Ibid., 53.
54. Veroff et al., *The Inner American . . . ,* 528.

Index